Nursing: Human Science and Human Care

A Theory of Nursing

Nursing:
Human Science
and Human Care
A Theory of Nursing

Jean Watson, RN, PhD, FAAN

Professor and Dean
School of Nursing
Director
Center for Human Caring
University of Colorado Health Sciences Center
Denver, Colorado

Pub. No. 15-2236

National League for Nursing • New York

18110569

ISBN 0-88737-417-4

Second printing 2M-1190-025950

Manufactured in the United States of America.

109/90

To Douglas, Jennifer, and Julie

CONTENTS

PREFACE

The aim of this book is to solve some conceptual and philosophical problems about nursing that exist in my mind. It is my hope that it will also guide others to join me in my quest to elucidate the human care process in nursing, preserve the concept of the person in our science, and better our contribution to society.

This writing process calls to mind a quote I saw in Bombay, India that was framed and hung on the bedroom wall of the children's room in a physician's home I was visiting. It read: "Life is not a problem to be solved, but a mystery to be lived." This saying expresses some of my own conceptual conflicts about nursing. Because the practice of nursing is in itself a mystery to be lived, some of nursing's problems are not ones that can necessarily be solved. Nevertheless, how can I, or we, go about highlighting "the mystery to be lived" well enough to have it valued, esteemed, "seen," developed, and incorporated into nursing education, practice, and research?

In setting forth my ideas about nursing and people in general, I found that having distance helps clarify my thoughts. As I began working on my manuscript, I found myself in Perth, Australia, some 4000 miles from home, overlooking the calm, peaceful Swan River, feeling the fresh sea breeze from the Indian Ocean.

Before sitting down to write I went through a stream of consciousness to create my own context for writing about the human-to-human care process of nursing. My free association included thoughts about people in general as well as diverse groups of people.

The human care process in nursing is, I believe, connected to other human struggles and to the tearing and wounding that can happen to a person or a race, a culture, or a civilization. This intensely human process of nursing can be a struggle for the professional nurse during a time of scientism and high technology.

In my first book, *Nursing: The Philosophy and Science of Caring*, I had no inclination to refer to my ideas as a theory. That earlier work came forth in my attempt to solve some conceptual and empirical problems about nursing, what comprises nursing, and how various components of nursing relate to and direct education, practice and research. *Nursing: The Philosophy and Science of Caring* was, in fact, a treatise on nursing.

When formulating my ideas, I began to structure a set of beliefs and concepts and to organize a body of knowledge and principles underlying human behavior in health and illness. Through this process came the ten "carative" factors in nursing. While my work was not a scientific theory per se, I was indeed theorizing about nursing and consequently going through the beginning stages of developing my own "theory" of nursing.

The present book is an extension of my previous work on caring. My ideas for this volume have evolved over the years, and have become crystallized during the past five years in my work with doctoral students. Bits and pieces of my ideas have gradually become fused into what I call transpersonal caring. I have directly and freely drawn upon the ideas of Carl Rogers in the definition of the self, but I have also been directed by my own values and beliefs about "the person" and life which are reflected in the inclusion of the soul as an important force in my concept of the person. My orientation is clearly "phenomenological-existential" and spiritual. One can also find in my ideas concepts associated with Hegel, Marcel, Whitehead, Kierkegaard and most certainly Eastern philosophy.

Sally Gadow's work on existential advocacy was also a source of influence and inspiration in refining and validating my ideas. Her concepts on moral ideal, intersubjectivity and human dignity have helped to ground my work. The work was also inspired and enhanced by my international travel and experiences during my sabbatical (research and study leave) from 1981 to 1983. My journeys included profound cultural-spiritual encounters in New Zealand, Australia, Indonesia, Malaysia, The Republic of China, Thailand, India and Egypt.

While this work is primarily intended for graduate students in nursing, it is also well suited to faculty members and to students of upper division baccalaureate programs. Hopefully, it will also touch a need in professional nurses who are struggling with the day-to-day world of human care in nursing.

I hope this work will provide greater clarity about my perspective on and approach to nursing. In addition, I hope this book will help to maintain and elevate the standard of the nursing profession and improve the welfare of people receiving and delivering nursing care. Finally, I hope the ideas contained in this volume will stimulate additional theory development by reaching out and touching the human mind and heart.

Jean Watson

ACKNOWLEDGMENTS

Acknowledgments go to all the nursing doctoral students at the University of Colorado from 1978 to 1984 who believed in nursing's "new" science and to Glenn Webster whose approach to the history and philosophy of science inspired me to go beyond the usual limits and "be metaphysical."

I also wish to recognize Margie Martin of New Zealand and Papua New Guinea, Gay Bernard of Kalgoorlie, Western Australia both of whom idealized my ideas of transpersonal caring in their ways of "being-in-the-world," and Annie and Angela, the community field nurses at the Aboriginal mission in Cundeelee for their special caring practices. Of course, I also have to include Bill Wesley, the magnificent tribal Aborigine who taught me about his people and their dreamtimes and whose dignity inspired my soul.

I certainly thank Mr. Charles Bollinger of Appleton-Century-Crofts for his gracious tolerance and patience. My sincere appreciation goes to Jack and Loretta Hempstead for their quality assistance in deciphering and typing the manuscript. My acknowledgments and appreciation also go to the Kellogg Foundation, the Centre for Advanced Studies, Division of Health Sciences, Western Australian Institute of Technology and the National Science Foundation of National Taiwan University for their support and assistance during the sabbatical that allowed me to work on this volume while I was a Visiting Kellogg Fellow in Australia.

My special, loving gratitude goes to Douglas, who encouraged, inspired, supported and critiqued my ideas and contributed to their development.

Lastly, I deeply value the joy of Jennifer and Julie whose spirit is ever with me.

1

INTRODUCTION:
Context For Theory Development

And the language of science cannot be freed from ambiguity, any more than poetry can;—in spite of its tidy look, the structure of science is no more exact, in any ultimate and final sense, than that of poetry.

J. Bronowski
The Identity of Man

This work represents the development of a nursing theory—not a hard scientific theory that is absolute, verifiable, quantifiable, and rigidly testable and that leads to facts, truths, and axiomatized statements—but nevertheless, a theory. It is a theory because it helps me "to see" more broadly (clearly), and it may be useful in solving some conceptual and empirical problems in nursing, and in human sciences generally. These views are presented with the hope that they may help others to see, to view phenomena in a new or different way, perhaps to develop or to attempt a new starting point, to use a new lens when focusing on the phenomena of human behavior in health and illness. Ideas are proposed about how nursing connects with and serves people, which, in turn, advances society's knowledge of human conditions and advances nursing's contribution to the welfare of society.

DEFINITION OF THEORY

A theory is an imaginative grouping of knowledge, ideas and experience that are represented symbolically and seek to illuminate a given phenomenon. "Science," scientific development, and theory development are all related to art, the humanities, and philosophy. All are related to imagination, creativity, mystery, personal problem-solving, and attempts at "being" in relation to the universe. An artist is as scientific as a scientist is artistic. I reject methods that ascribe an increasing degree of reality to numbers and factual information, when at the same time the human need for esthetics, wholeness, faith, inspiration, and a sense of wonder, mystery, and discovery is pushed farther into the background.

I reject definitions and interpretations of science and scientific inquiry that bury the quest for discovery, beauty, creativity, and a higher sense of being-in-the-world. I want nursing to move beyond objectivism, verification, rigid operations, and definitions and concern itself more with meaning, relationships, context, and patterns. I want nursing to be more concerned with the pursuit of hidden truths and new insights, developments of new knowledge in relation to human behavior in health and illness, and to make new discoveries of how to be in a professional human caring relationship with individuals to serve society. As Lauden states,[1]

> Many of the theories within any evolving research tradition will be mutually inconsistent rivals, precisely because some theories represent attempts, within the framework of the traditions to improve and connect with their predecessors.
>
> A research tradition is a set of general assumptions about the entities and processes in a domain of study and about the appropriate methods to be used for investigating the problems and constructing the theories in the domain. A successful research tradition is one which leads, via its component theories, to the adequate solution of an ever increasing range of empirical and conceptual problems.

The broader task underlying this work is to help clarify the nature of nursing's contribution to humankind by way of a theoretical perspective that helps to solve some empirical and conceptual problems related to nursing science and the preservation of the person and human care. Nursing's challenge of the day is to break the old bonds of preoccupation with procedures, facts per se, rigid definitions, strict rationalism, operationism, variable manipulation, and so forth. We must recognize other ways of knowing and alternative views of science. We need additional approaches to study and research the area of human health–illness experiences and human care that do not exhaust the meaning of the facts or concepts. We need to develop methods that retain the human context and allow for advancement of knowledge about the lived world of human experience.

My position is even stronger because I view nursing as both a human science and an art, and as such it cannot be considered qualitatively continuous with traditional, reductionistic, scientific methodology.

Figure 1.   Horizontal concept continuum.

Concepts

Since *concepts* are the basic building blocks for any theory, it is helpful to examine different ways in which concepts are treated by a theorist as a starting point for theory development.

Concepts and definitional terms in a given theory may extend on a horizontal continuum from the very absolute or concrete to the very relative or abstract (Fig. 1). A theorist's starting point on the concrete–abstract continuum will influence further directions for the theory. Any theory's structure, flexibility, utility, and so on, will be influenced by how the basic concepts are treated in the beginning stages; all other aspects of the work will flow from that starting point. Let's take a concept as elusive as nursing. The way the concept is treated may range from a very absolute, specific, concrete set of actions, tasks, and behaviors that one can directly observe and measure to a very abstract, dynamic, philosophical ideal of nursing that is relativistic and suggests a constantly changing process with different meanings, and emotional associations. Moreover, nursing may be codefined by the patient. If a concept were placed somewhere in the middle of the concrete–abstract continuum, it may suggest a balanced view of the concept under question. In other words, the concept of nursing may be treated in such a way that abstractions are allowed for, but abstract notions represent empirical reality to a degree that is comprehensible by others, both in philosophic, intellectual terms, and in some comprehensive form in the world of practice.

If the concept of nursing is placed in the middle of the continuum, for example, nursing may be defined in such a way that it incorporates the action, doing, behavioral aspects of nursing, but also allows for some relativistic notions such as the meaning that nursing may have to the experiencing patient. It may allow for the nursing presence to exist in a patient's mind, even if the nurse is not present physically. Such a balanced development of the concept of nursing would provide both for objective and subjective dimensions of nursing with some specific measurable, observable aspects, but also some abstract, theoretical aspects that may or may not be measurable in behavioral terms, but may be understandable and meaningful within a theory.

In addition to the absolute–relative continuum as a starting point, another continuum can be vertically superimposed. This can be referred to as a vertical static-dynamic continuum (Fig. 2). For example, could

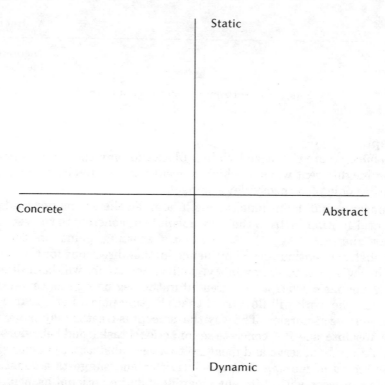

Figure 2. Horizontal and vertical concept continua. *(Vertical continuum idea attributed to Dr. Glenn Webster.)*

the way in which a concept is defined result in some external laws, universal truths and certainties? If so, that may be useful, but there is also the danger that they may become static and deadened after verification. On the other hand, if the concept is viewed as dynamic, temporal, changing, and unfinished, the expression of the concept may allow for diversity, lack of certainty, and evolution.

These starting points on how the beginning concepts of any theory are expressed and defined will determine where the theory ends up and to what extent it is useful in describing, explaining, understanding, and seeing the phenomena in question. Furthermore, the starting point on the concrete–abstract, static–dynamic continuum will also determine what methodologies are suitable for further scientific work, further theory development, and research.

If the starting point for concept expression is at the concrete end of the horizontal continuum and the static point on the vertical continuum then the scientific perspective of the theory would be one of verification, absolutes, and measurables, including acceptance of truths, laws, and facts once they are tested. If a concept is placed on the abstract–dynamic ends of the two continua, then the scientific perspective would be consistent with a view of science as discovery rather than verification, a pursuit of meaning and understanding that may be less than true but may hold promise for new ways of knowing, seeing, or answering questions, albeit tentative.

The *verification–acceptance* method of science is more concerned with science as a finished product. The *discovery–pursuit* method of science is more concerned with science as a process that is continuous, ongoing, and unfinished. To allow for movement of a concept on the vertical/horizontal axis, it is more useful to avoid being locked into either end of the two continua. However, depending upon one's approach to science (as a product to be verified and accepted, or as a process to be discovered and pursued), a theorist may end up on different points on the two continua. The theorist's orientation toward science will influence his or her starting point for developing and defining the specific concepts that become the foundation and building blocks of the theory. In turn, the starting point for developing the concepts will determine the methodologies and approaches to theory development, theory structure, and theory use. Thus the starting point will determine the end. The clearer one is about the starting point, the more one can see where or why one may end up in a certain place. One can also begin to recognize why a given theory's use was limited if the method for further work was inconsistent with the starting point. Some theories, for example, are investigated by verification methods related to science as a product when the theoretical concepts are more consistent with process discovery, pursuant to methods related to science as process. When that happens, the theory often dies. This has frequently occurred in periods of nursing theory developments.

There has been intellectual confusion in nursing as to which continua should operate; there have been some conceptual inconsistencies between and among some of the dimensions. Often there are indications that concepts should be open, fluid, changing, and consistent with human behavior and nursing science, but ideas have frequently been trapped by applications of rigid testability and methods of verification and acceptance that are not consistent with nursing as a human science. There has also been some bootlegging of nursing concepts into molds of other disciplines that are not adequate for nursing science. At the same time nursing is beginning to recognize that it can broaden its perspective on science, consistent with the changing nature of the history and philosophy of science, and explore numerous methods for advancing knowledge. There is a reawakening consciousness and confidence for nursing to pursue a nursing science consistent with its own heritage and tradition which is related to human science and human care.

Before proceeding I would like to place my own theoretical perspective on the continua so the reader knows my starting point (Fig. 3).

Consistent with my views of science, I am concerned with process,

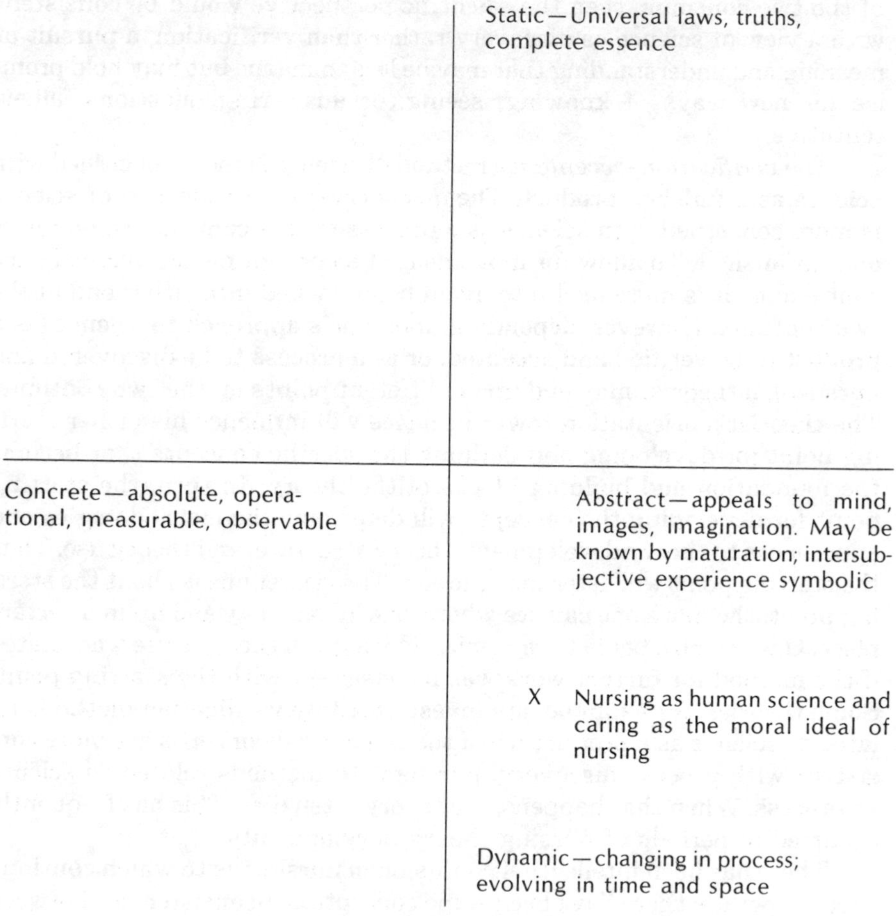

	Static—Universal laws, truths, complete essence
Concrete—absolute, operational, measurable, observable	Abstract—appeals to mind, images, imagination. May be known by maturation; intersubjective experience symbolic
	X Nursing as human science and caring as the moral ideal of nursing
	Dynamic—changing in process; evolving in time and space

Figure 3. Concept continua—concrete to abstract and static to dynamic.

the method of discovery, and the pursuit of hidden meanings in nature and life. Therefore, I would place my ideas toward the far right-hand side of the abstract end of the horizontal continua and toward the lower end of the vertical axis, associated with dynamic movement, lack of certainties. At the same time, however, I am interested in movement back and forth and up and down on the dimensions—but my starting point is *more toward the lower, right quandrant.*

Any of the phases of theory development, such as concept definitions, concept representation, formulation of propositions, checks with empirical world for meaning, clarification, may be more or less concrete–abstract, static–dynamic, depending on the theory and the theorist's world view and research tradition.

For example, do the concepts in the theory induce a mental representation or high level abstraction? A behavioral representation or a physical representation? Is the concept dynamic or time dependent? Is the focus micro or macro? Is the scope partial or comprehensive? Is the outcome more descriptive where abstractions are visualized and imagined, or reduced to concrete language and words and/or numbers that are devoid of fuller meanings, or restricted to only the operationalized definitions and observable events? Is the focus on what ought to be (idealism) or what is (realism)? Is there a high level of generalization at one level, but a lack of specific facts and details at a more concrete, individual level? (See Fig. 4.)

Consider the following illustrations and examples from nursing theory. Martha Rogers's theory of nursing and concept of energy field and Roy's concept of adaption. Rogers's concept of energy field is defined as electrical in nature in a continual state of flux and varies continuously in its intensity, density and extent.[2]

The concept is broad, abstract, theoretical, almost metaphysical; it has a great deal of generalization but is void of specific facts. It helps one to see and comprehend and grasp an essence without spelling out details of specific behavior; it evokes higher level mental images.

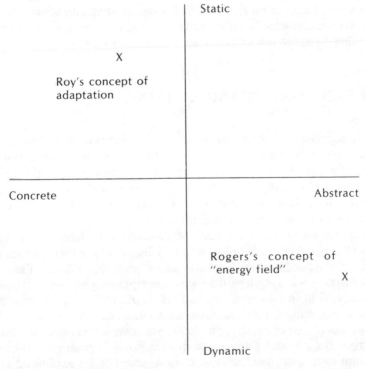

Figure 4. Placement of Roy's and Rogers's theory on concept continua.

Roy's concept of adaptation includes three classes of stimuli:[3] (1) focal stimuli, (2) contextual stimuli, and (3) residual stimuli. As such, adaptation results from a response to a stimulus. This concept is placed in the concrete–static one because there is a limited generalization to higher levels of mental images. The concept is limited to specific stimuli that are more detailed and factual, measureable entities, but void of high-level generalization. It may help one to see immediate presenting behavior, but does not help in grasping an essence or a higher abstract mental image, as it is concerned more with specific observable behavior of a given time.

These two examples illustrate a macro (comprehensive) perspective (Rogers's) and a more micro (partial) perspective (Roy's). Sometimes different dictates and orientations have been dismissed under the guise of a science–nonscience schism. Consequently, a perspective that is too global and too abstract, even if it is rich and powerful as a mental device is sometimes discredited. Likewise, depending upon one's approach to science (whether the concern is with verification and product rather than discovery and process), one's preference will be for one theory or the other. What is perhaps more useful in science, and certainly nursing, is: (1) to be clear about one's particular leanings, (2) to encourage a diversity of approaches for theory development in nursing, and (3) to select an approach consistent with one's own values and beliefs. Preferable also is a theory that allows for placement on the specificity–generalization scale and the concrete–abstract, static–dynamic continua that is consistent with the nature of nursing phenomena in question within the theory.

NEW LENS FOR SEEING NURSING

Nurse scientists and practitioners must treasure some of their nonlinearities and other unexpected results and avoid preconceptions based upon ingrained ideas. We need to move away from homogeneity of thinking and seek new breakthroughs, develop new ways of seeing the usual. Theories may need some isolation to evolve in order to prevent their dilution and submersion of the common and ordinary.

Nursing science has to work at changing its lens to see anew and appreciate some of its beauty, art, and humanity as well as its science. Perhaps the issue for nursing is to acknowledge that it is not like the traditional sciences—it requires its own description, possesses its own phenomena, and needs its own method for clarification of its own concepts and their meanings, relationships, and context.

The notion of causality, for example, cannot necessarily be carried over from the natural sciences to nursing and human care. The human-to-human caring transactions of nursing cannot be explained or understood with a positivistic, deterministic, materialistic mind set. "There seems to be a single starting point (for nursing), exactly as for all the other sciences; the world as we find it and live it,—naively, uncritically. The naivete may be lost as we proceed. Problems may be found which were at first completely hidden from our eyes. For their solution it may be necessary to devise concepts which seem to have little contact with direct primary experience. Nevertheless, the whole development must begin with a naive picture of the world."[4]

When one looks back over the history of nursing science, it has not had one sure sense of direction but several quite unsure directions and various research traditions. The apparent differences between nursing and other branches of science may result from the difficulty of the context and processes and concepts involved, such as nursing, caring, humans, life, human relationships, health, healing, and so on. It is also significant that many of the problems of nursing science, theory, and research have not been systematically attacked in a scientific context and are just now being explored. Recently, a growing body of empirical information has emerged, and there has been serious concern expressed over methodological issues. A variety of efforts are being pursued to bring a selected body of knowledge into a framework for studying, researching, and practicing nursing.

The features of traditional science and medicine that make it anomalous to nursing science have been stated in different ways, but can be summed up in three characteristics—objectivism, scientism, and technism. These are shared by the two most influential philosophical systems in the history of science, Cartesianism and Positivism.

Nursing is undergoing a questioning process as to whether it should continue to align itself with traditional science to improve practice or to abandon science in favor of some other approach to reality. It would be inappropriate and irresponsible to discard science and discount scientific progress, but at the same time, science, as traditionally viewed and presented, must be questioned and challenged in nursing and human sciences.

It is appropriate that nurses question the impersonal, objective model of science for the personal unique and gestalt experiences. The science paradigm for nursing must allow human phenomena to emerge and still be investigated. The method must be such that the humanness of a relationship between two beings is neither diminished nor lost.

Without wanting to create an additional set of false dichotomies between science and art, or between traditional–medical–natural science models and human science, and nursing models, Table 1 does set forth some of the major differences that help me to clarify nursing's scientific context.

Table 1 is not intended to reinforce differences but rather to heighten awareness of different assumptions. These differences direct our thinking and behavior and need to be examined continually. Sometimes the differences can be reconciled through a dialectical process of thesis, anti-

TABLE 1. DIFFERING PERSPECTIVES BETWEEN TRADITIONAL SCIENCE AND HUMAN SCIENCE

Traditional Medical Natural Science Context	Emerging Alternative Nursing Human Science/Context for Caring
Normative	Ipsative
Reductionistic	Transactional
Mechanistic	Metaphysical Humanistic—contextual
Method centered	Phenomena centered
Neutrality of values	Value laden; values acknowledged, clarified
Disease centered on pathology–physiology, the physical body	Person–experience centered Human responses to illness and personal meanings of human condition
Ethics of "science"	Human–social ethics–morality
More quantitative	More qualitative
Absolutes, givens, laws	Relativism, probabilism
Human as object	Human as subject
Objective experiences	Subjective–intersubjective experiences
Facts	Experience, meaning
Nomothetic	Idiographic + /nomethetic
Concrete—observable	Abstract—may or may not "be seen"
Analytical	Dialectical, philosophical, metaphysical
Science as product	Science as creative process of discovery
Human = sum of parts ex. (bio-psycho-socio-cultural-spiritual-being)	Human = mind/body/spirit gestalt of whole being (not only more than sum of parts, but different)
Physical, materialistic	Existential—phenomenological–spiritual
"Real" is that which is measurable, observable, and knowable	"Real" is abstract, largely subjective as well as objective, but it may or may not ever be fully known, observable, fully measured, what is "real," holds mystery and unknowns yet to be discovered

thesis, and synthesis. Some remain as differences and need to be at least acknowledged since they represent different starting points and lead us in different directions.

As a way of summarizing my ideas within a context of theory development, Chapter 9 later provides a synopsis and a structural overview of my theory of human care. It also offers the reader a succinct analysis of the theory and its major components, including subject matter, values, goals, agents of change, interventions, perspective, context, approach, and method.

REFERENCES

1. Lauden, L. Progress and Its Problems: Toward a Theory of Scientific Growth. Berkeley, Calif.: University of California Press, 1977, 81, 82, 87.
2. Rogers, M.E. An Introduction to the Theoretical Basis of Nursing. Philadelphia: Davis, 1970, 90, 91, 92, 104, 113.
3. Roy, Sister C. Introduction to Nursing: An Adaption Model. Englewood Cliffs, N.J.: Prentice-Hall, 1976, 22, 30-32, 38.
4. Kohler, W. Gestalt Psychology: An Introduction to New Concepts in Psychology. New York: Liveright, 1947, 3-4.

BIBLIOGRAPHY

Abdellah, F.G. The nature of nursing science. Nursing Research, 1969, 18, 390.

Alexandersson, C. Amedeo Giorgi's empirical phenomenology (Publication No. 3). Swedish Council for Research in Humanities and Social Sciences, Department of Education, University of Goteborg, Sweden, 1981.

Dennis, N. New Methods for Research. Paper presented at Western Australian Institute of Technology, Western Australia, May 1982.

Flaskerud, J.H. & Halloran, E. Areas of agreement in nursing theory development. Advances in Nursing Science, 1980, 3(1).

Gaylin, W. Caring. New York: Knopf, 1976.

Giorgi, A. Psychology as a Human Science. New York: Harper & Row, Pub., 1970.

Hall, L.E. Nursing—What is it? Canadian Nurse, 1964, 60, 150-154.

Henderson, V. The nature of nursing. American Journal of Nursing, 1964, 64, 62-68.

Hyde, A. The Phenomenon of Caring Part I-Part IV, (Vols. 10-12). American Nursing Foundation, 1975-1977.

Johnson D. State of art of theory development in nursing. In National League for Nursing (Ed.), Theory Development: What, Why, and How. New York: National League for Nursing, 1978.

Johnson, R. In Quest of a New Psychology. New York: Human Sciences Press, 1975.

King, I. Toward a Theory for Nursing. New York: Wiley, 1971.

Koch, S. Psychology cannot be a coherent science. Psychology Today, 1969, 3(4), 64, 66.

Koch, S. (Ed.). Psychology: A Study of Science. New York: McGraw-Hill, 1959.

Kohler, W. Gestalt Psychology. New York: Liveright, 1947.

Kreuter, F.R. What is good nursing care? Nursing Outlook, 1957, 5, 302-305.

Lauden, L. Progress and Its Problems: Toward a Theory of Scientific Growth. Berkeley, Calif.: University of California Press, 1977.

Leininger, M. (Ed.). Caring. Thorofare, N.J.: Charles B. Slack, 1981.

Leininger, M. Conference on the nature of science in nursing. Introduction: Nature of science in nursing. Nursing Research, 1969, 18(5).

Leininger, M. Foreword. In J. Watson (Ed.). Nursing: The Philosophy and Science of Caring. Boston: Little, Brown, 1979.

Levin, M. Holistic nursing. Nursing Clinics of North America, 1971, 6(2).

Marton, F. Describing conceptions of the world around us. Reports from Institute of Education, University of Goteborg, Sweden, 1978.

Marton, F. & Svensson, L. Conceptions of research in student learning. Higher Education, 1979, 8.

Mayerhoff, M. On Caring. New York: Harper & Row, Pub., 1971.

Murphy, J. (Ed.). Theoretical Issues in Professional Nursing. New York: Appleton-Century-Crofts, 1971.

Newman, M. Theory Development in Nursing. Philadelphia: Davis, 1979.

Nightingale, F. Notes on Nursing: What It Is and What It Is Not. New York: Appleton, 1860.

Norris, C. M. (Ed.). Proceedings, First Nursing Theory Conference. University of Kansas Medical Center, Department of Nursing Education, March 20–21, 1969. Kansas City, Kan.: University of Kansas Press, 1969.

Oiler, C. The phenomenological approach in nursing research. Nursing Research, 1982, 31(3), 178–181.

Omery, A. Phenomenology: A method for nursing research. Advances in Nursing Science, 1982, 5(2), 49–63.

Parse, R. R. Man–Living Health: A Theory of Nursing. New York: Wiley, 1981.

Paterson, J. D. & Zderak, L. Y. Humanistic Nursing. New York: Wiley, 1976.

Peplau, H. Interpersonal Relations in Nursing. New York: Putnam: 1952.

Pickering, M. Introduction to qualitative research methodology. Paper presented at the American Speech–Language and Hearing Association, Detroit, November, 1980.

Rist, R. C. On the relations among educational research paradigms: From disdain to detente. Anthropology and Educational Quarterly, 1977, 8, 42–49.

Rogers, M. Theoretical Basis of Nursing. Philadelphia: Davis, 1970.

Roy, Sister C. Introduction to Nursing: An Adaptation Model. Englewood Cliffs, N.J.: Prentice-Hall, 1976.

Stevens, B. Nursing Theory. Boston: Little, Brown, 1979.

Spicker, S. & Gadow, S. (Eds.), Nursing Images and Ideals. New York: Springer, 1980.

Valle, R. & King, M. Existential Phenomenological Alternatives for Psychology. New York: Oxford University Press, 1978.

Van Kaam, A. L. Phenomenological analysis: Exemplified by a study of the experience of being really understood. Individual Psychology, 1959, 15, 66–72.

Watson J. Nursing: The Philosophy and Science of Caring. Boston: Little, Brown, 1979.

Watson, J. Nursing's scientific quest. Nursing Outlook, 1981, 29(7), 413–416.

Watson J. Professional identity crisis—Is nursing finally growing up? American Journal of Nursing, 1981, 81 1488–1490.

Watson, J. Reflections of new methodologies for study of human care. In M. Leininger, (Ed.), Qualitative Methodologies in Nursing. New York: Grune & Stratton, 1984 (in press).

Yura, K., & Torres, G. Today's conceptual frameworks within baccalaureate nursing programs. National League for Nursing Pub. No. 15–1558. New York: National League for Nursing, 1975, 17–25.

2

NURSING AS HUMAN SCIENCE

To effectively interpret the truly great role that has been assigned her, neither a liberal education nor a high degree of technical skill will suffice. The nurse must also be master of two tongues, the tongue of science and that of the people.

Annie Warburton Goodrich *

In some of my earlier writings, I tried to present a case for nursing as an art and a science, making it more akin to a human science tradition and what I called the science of caring. A second position I have taken is that nursing has submerged both its scientific and artistic heritage in its scientific quest.

Such a position about the nature of nursing and nursing science is hardly unique. In fact an exploration of the early writings on nursing shows similar ideas about the nature of nursing.[1]

Even though nursing is still evolving and has yet to actualize the ideas and ideals of the early nursing leaders, contemporary theories of nursing still generally promote similar ideas about nursing.[2] As nursing progresses, it is breaking away from the traditional medical–scientific bondage and attending to developing its own scientific heritage.

---

* Taken from a reprint of addresses, The Social and Ethical Significance of Nursing published in 1973 by Yale University School of Nursing, New Haven, Conn. Originally published by Redbook Magazine, 1929.

It appears that early nursing leaders were attempting to create a research tradition for nursing; that is, they tried to establish a set of general assumptions about entities and processes in the domain of study and the appropriate methods to be used for investigating problems and constructing theories.[3] Florence Nightingale, for example, talked of a "new art and a new science" and presented nursing as an act that required organized and scientific training.[4] She did not create a false dichotomy between science and art.

Later leaders such as Virginia Henderson, Lydia Hall, and Frances Krueter promoted concepts of nursing that were consistent with Nightingale's. Henderson, for example, defined the nurse's role as very subjective and qualitative. She believed the nurse should " . . . get inside the skin of each of (her) patients in order to know what (he/she) needs."[5]

According to Annie Goodrich, nursing "imbues the simplest acts with importance and instills a desire for the upmost skill and accuracy in their performance. It commands beyond all other drives devoted service, . . . a broad perspective, . . . rigorous analyses, close association with scientific findings, fine perceptions, (and) enduring tolerance born of understanding."[6]

The common and broad themes from nursing's heritage about the nature of nursing that I adopt are:

1. A view of the human as a valued person in and of him- or herself to be cared for, respected, nurtured, understood and assisted; in general a philosophical view of a person as a fully functional integrated self. The human is viewed as greater than, and different from, the sum of his or her parts.
2. An emphasis on the human relationship and transaction between persons and their environment and how that affects health and healing in a broad sense.
3. An emphasis on the human-to-human care transaction between the nurse and person and how that affects health and healing.
4. An emphasis on the non-medical processes of human care and the nurse's caring for persons with various health–illness experiences.
5. A concern for health; the promotion of health and well-being.
6. A position that nursing human care knowledge is distinct from, but complementary to, medical knowledge.

The problem today seems to be that nursing has not developed the *science* of nursing in accordance with its *theories*. Furthermore, there are conflicting paradigms operating among different views of nursing theory, nursing practice, and nursing research. This conflict is related to the fact that the early views of nursing that have persisted across time have still not been realized in education or in practice. The intervening years between the early leaders and contemporary theorists have been fraught with nursing's struggle to advance as a discipline and a profession. We have been caught between the paradigm of medical science with a body-part view of a person and the pardigm of natural, hard science with an emphasis on unsurpassable control, rigor, objectivism, neutrality of value, facts, procedures, skills, technology, and so on.

Consequently, nursing still has a long way to go in adopting a meaningful philosophical foundation for its theories and its science that is consistent with past and present visions, images, and ideals of nursing leaders.

HUMAN SCIENCE

The notion of a "human science" is a term used by Giorgi[7] in his attempt to describe psychology as a discipline committed to the study of the person as a whole, as opposed to the psychoanalytic or behaviorist views of psychology. This dilemma also applies to nursing's historical and traditional ties to medical science and to the psychoanalytic and behaviorist views of the person. This approach has been compounded with the natural science, reductionist research methods nursing has adopted for its science. At the same time nursing has a strong commitment to care of the whole person and a concern for the health of individuals and groups of persons.

Recently, ideas consistent with a human science emphasis for nursing science and research have been advanced in the nursing literature by such nursing theorists, researchers, and authors as Davis,[8] Watson,[9-11] Winstead-Fry,[12] Parse,[13] Webster, Jacox, and Baldwin,[14] Downs,[15] Munhall,[16] Chinn,[17] Donaldson,[18] Newman,[19] Leininger,[20] and others.

In general, if one adheres to a human science perspective the following areas are acknowledged:

1. There is an anomaly between the organismic concept of person in medicine and traditional psychology and the concept of person as a whole referred to in nursing.
2. There is a strain between the study of person as a *whole* (and human responses) and the process of nursing care and the traditional reductionistic assumptions of natural science, basic sciences, and biomedical sciences.
3. Nursing is a relatively young discipline, dating back to the mid- to late-nineteenth century: hence, it is susceptible to the temptation to follow the rule of the older natural sciences, without raising important philosophical, epistemological, ethical, and scientific questions relevant to the study of nursing and nursing phenomena.

Other philosophical and conceptual aspects related to a human science context for nursing study are:

1. Nursing views human beings as experiencing subjects.
2. There is an interconnected evolution of the human and the world.
3. Health is a process.
4. Change is ongoing; nurse and person are coparticipants.

A human science perspective allows nursing scholars and researchers to raise serious questions about nursing science and the new directions that nursing must take to be true to its subject matter and its social and scientific responsibility. Moreover, a human science perspective opens new vistas and new possibilities for humans and their world of health–illness experiences. Such a perspective allows for the questioning of ultimate meanings and ethical values of humans, health and nursing.

In summary, a human science context is based upon:*

- A philosophy of human freedom, choice, responsibility.
- A biology and psychology of holism (nonreducible persons interconnected with others and nature).
- An epistemology that allows not only for empirics, but for advancement of esthetics, ethical values, intuition, and process discovery.
- An ontology of time *and* space.
- A context of interhuman events, processes, and relationships.
- A scientific world view that is open.

It is critical at this point in time that nursing adheres to a perspective that does not disengage nursing's ultimate meanings, intuitions, and relevance from its esthetics, ethics, science, and practice.[21]

In our past attempts to be scientific and advance as a profession and a discipline, we have suddenly reached a junction that can lead us in two different directions. One path is that of traditional medical science with its distinctive epistemology. The other path acknowledges nursing as a human science with another epistemology.

The traditional science approach is to take the concepts, viewpoints, and techniques of natural science and medicine and apply them to nursing and the lived-in world of human health–illness experiences. To do so, however, we are making certain assumptions about human life and the human caring process in nursing that are nonhuman in character, that is, adopting:

---

*This section draws from the author's previous work: Watson, J. Reflections on new methodologies for study of humancare. In M. Leininger, (Ed.), *Qualitative Methodologies in Nursing.* New York: Grune & Stratton, 1984 (in press).

- a distinctive epistemology of empirics
- a philosophy of human determinism
- a biology and psychology of organismic–mechanistic physicalism
- an ontology of space versus time
- a context of parts, with mind, body, and spirit splits
- a scientific world view that is closed
- a methodology of analysis and validation of repetitive facts.[22]

This traditional approach is one in which we accept the position that our fundamental world view is (scientifically) settled and the major responsibility of contemporary nurse practitioners, scholars, and researchers is simply to add to the increasingly complex store of knowledge. This path is haunted by the restricted thinking of the medical paradigm that labels, categorizes, manipulates, controls, and treats disease; if not disease, patients, if not patients, variables.

Such an approach is marred by medical values, goals, and interventions laden with paternalistic notions that are inconsistent with the nursing human care and the notion of a person as an end in and of him- or herself. If the discipline of nursing operates under the traditional approach it adopts the ethic of empirical science. This ethic of science is recognized by nursing's adherence to a research tradition that concentrates on objectivity, facts, measurement of smaller and smaller parts, and issues of instrumentality, reliability, validity, and operationalization to the extent that nursing is in danger of exhausting the meaning, relevance, and understanding of the values, goals, and actions that it espouses in its heritage and ideals. Such a position can disengage nursing's ultimate meanings and intuitions from its esthetics, ethics, science, and practice. It is limited by its starting point and fundamental scientific and philosophical restrictions of human life.

On the other hand, if we view nursing as a human science, we can combine and integrate the science with the beauty, art, ethics, and esthetics of the human-to-human care process in nursing. Human science is based upon an epistemology that can include metaphysics as well as esthetics, the humanities, art, and empirics.

Nursing can be discovered anew with a more meaningful philosophical foundation based on human rather than nonhuman values—returning to our roots in order to find and establish our own foundation for the future.

As a human rather than a traditional science, nursing can view human life as a gift to be cherished—a process of wonder, awe, and mystery. Nurses can choose methods that allow for the subjective, inner world of personal meanings of the nurse and the other person. We can choose to study the inner world of experiences rather than outer world of observation. We can choose to be a part of our method and involved in the clinical research process rather than be distant, objectively remote, and primarily concerned with the product of science. We can choose to pursue more of the private, intimate world of human care and inner subjective human experiences, rather than to concentrate on the public world of nonhuman cure techniques and outer behavior.

This different path can expand our limited thinking and allow us, as professionals and scholars, to develop new pictures of what it means to be human, to be a nurse, to be ill, to be healed, and to give and receive human care. Whether we, as nurses, see human life one way or the other is the result of different intentional acts; moreover, how we choose leads us in very different directions and has very different consequences for our practice, our science *and* our methodology. Indeed, our choice has consequences for nursing's contribution to society in the preservation of humanity.

Perhaps nursing can learn from the wisdom and experience of Sigmund Koch. Sigmund Koch was an eminent theoretical scholar in psychology and the author of *Psychology: A Study of Science.*[23] After a 30-year career devoted to the exploration of the conditions for psychology becoming a scientific enterprise, he concluded: "Psychology cannot be a coherent science and the end result of the enterprise has been nothing more than a prolification of pseudo-knowledge." Koch further concluded that "after 100 years of experience, psychology has failed and is lost and it must be discovered anew wherein it is established on a more meaningful philosophical foundation."[24]

Nursing is in danger of a similar demise if we do not seriously question our scientific approach. Nursing today has room for more optimism if we proceed with caution and stop emulating the natural sciences. We still have time to rediscover our commitments to humans and the human care process and establish a more meaningful philosophical foundation from which to proceed.

These new foundations for nursing are grounded in a professional human care process that connects with and becomes a part of the lived world of human experiences associated with health and illness. Nursing must question the position that the fundamental world view is settled and help to create a new world view of our subject matter, the human nature of our science, practice, and develop methods that are credible, meaningful, and true to our human caring phenomena.

NEW DIRECTIONS FOR THE FUTURE

More and more nursing theories and researchers are now beginning to posit a new nursing research tradition that is giving rise to what Kuhn refers to as a revolution. There is greater acknowledgment and public recognition that continued adherence to a medical model for nursing practice and adherence to a traditional natural science model for nursing science is not adequate for addressing the phenomena of human care in nursing and human responses to actual or potential health problems.

For perhaps the first time in its scientific development, nursing has recently had the opportunity to explore its own heritage, become recommitted to nursing values, goals, and philosophies, and explore research methods and options consistent with prevailing views about the nature of nursing. In so doing, nursing scholars and clinicians have begun to admit openly that an inconsistency or anomaly has existed and continues to exist between the medical tradition and/or natural science paradigm and the nature of nursing. Furthermore, such a dilemma has led nursing researchers and theorists to pursue more actively alternative approaches to developing nursing science. Advanced study in nursing is providing nursing scholars an opportunity to set forth a paradigm about persons-health-environment and human care processes of nursing different from the medical and natural science paradigm. In addition, nursing is investigating alternative methods for researching nursing phenomena.

As nursing science begins to make a paradigmatic shift for the study of nursing phenomena, it can be argued that there are some major differences in various assumptions underlying sciences. Table 2 provides different assumptions about the nature of reality, the nature of the inquirer-subject-object relationship, and the nature of truth statments.

These differences have been identified by Giorgi,[25] Marton and Svensson,[26] Marton,[27] Cook and Reichardt,[28] Watson,[29] and others, and hold much promise for nursing science. These differences will be acknowledged throughout this book.

The traditional science medical paradigm and the human science/nursing paradigm differ as shown in Tables 2 and 3.

Cook and Reichardt[30] characterize such paradigmatic differences as methodologically linked, but should not be viewed as mutually exclusive. Table 4 lists different attributes of the qualitative and quantitative paradigms.

Another way of viewing the different paradigms is to consider the various dimensions in a paradigm as a continuum, wherein there is movement away from, or toward, one end of the continuum to the other, depending upon one's assumptions, views of science, the nature of the phenomena under study, and so on. Table 5 represents the movement of various dimensions from one stand toward another, without proposing that they are mutually exclusive and always antagonistic. Table 5 does, however, indicate that as nursing seeks to preserve the concept and context of person, nurse, and human care, as part of its scientific phenomena, there will tend to be movement toward the qualitative or combination approaches in developing nursing as a human science.

The issue in nursing science of the present and the future is beyond method and paradigm per se, but rather is based upon one's world view

TABLE 2. ASSUMPTIONS UNDERLYING NURSING SCIENCE

Traditional Medical Science	Assumptions	Human Science
Quantitative—Rationalistic Inquiry		*Qualitative—Phenomenological Naturalistic Inquiry*
A single reality; parts separable, manipulable; variables independent context neutral or free	Nature of reality	Multiple realities; parts interrelated; variables dependent; context dependent
Inquirer-investigator can maintain distance and independence	Nature of inquirer-subject relationship	Inquirer and respondent interrelated; the inquiring process has an influence on the inquiry
Generalizations possible; enduring truth statements focus on nomothetic (group norm) knowledge; concern with similarities	Nature of truth statements	Full generalizations not possible; what results are working hypotheses for a particular context; focus on idiographic (individual) knowledge or combination; (individual) concern with difference

(Adapted from Guba, E.G. & Lincoln, Y.S. Epistemological and methodological bases of naturalistic inquiry. Educational Communication and Technology Journal, 1982, 30, 2-7.)

TABLE 3. CONCEPTIONS OF EIGHT ASPECTS OF SCIENCE AND DIFFERENCES BETWEEN TRADITIONAL SCIENCE PARADIGM AND HUMAN SCIENCE PARADIGM

	Traditional Science Medical Paradigm	Human Science Nursing Paradigm
Perspective	Objectivity, observational—measurable	Experiential—subjective, metaphysical
Description	Quantitative	Qualitative or combination qualitative/ quantitative
Conceptualization	Generalizable	Contextual
Relations	External—often statistically inferred	Internal—person confirmed
Comprehensions	Explanation—prediction	Understanding
Emphasis	Facts—data	Meaning
Use	Technical, validation of knowledge, extension of existing knowledge	Emancipatory—(new insights, theory, discovery, new knowledge)
Structure	Paradigm adherence	Paradigm transcending

(Adapted from Marton, F. & Svensson, L. (1979). "Conceptions of research in student learning," Higher Education 9:484. Elsevier Science Publishers. B.V., Amsterdam.)

and predisposes one to see the world and the events within it in profoundly differing ways.

In brief, the prevailing world view of nursing science, is becoming more and more phenomenologically inductive, subjective, process-oriented, and even metaphysical. Because nurses see the world in different ways from the medical–natural science paradigm, nursing theorists and researchers use different methods of inquiry. The differing methods are often translated into the qualitative and quantitative paradigm. Although these two paradigms are not mutually exclusive, they do represent two different ends of a continuum.

As nursing science begins to shift its emphasis and transcend existing paradigms, there are different lenses a nurse can use to approach the perception of nursing as a human science.

Some of the critical epistemological, scientific, ethical, esthetic, and methodological questions nursing has yet to explore are related to the nature of the human subject matter of the nurse and the person; how are they defined or codefined? What conditions facilitate and sustain the person as an end, not a means to some scientific or medical end? What conditions sustain human caring in instances of threatened humanity, biological or otherwise?

TABLE 4. ATTRIBUTES OF THE QUALITATIVE AND QUANTITATIVE PARADIGMS

Qualitative Paradigm	Quantitative Paradigm
Advocates the use of qualitative methods	Advocates the use of quantitative methods
Phenomonologism and verstehen: "concerned with understanding human behavior from the actor's own "frame of reference."	Logical-positivism: "seeks the facts or causes of social phenomena with little regard for the subjective states of individuals."
Naturalistic and uncontrolled observation	Obtrusive and controlled measurement
Subjective	Objective
Close to the data; the insider perspective	Removed from the data; the outsider perspective
Grounded, discovery-oriented, exploratory, expansionist, descriptive, and inductive	Ungrounded, verification-oriented, confirmatory, reductionist, inferential, and hypotheticodeductive.
Process-oriented	Outcome-oriented
Valid: real, rich, and deep data.	Reliable: hard and replicable data
Ungeneralizable; single case studies	Generalizable; multiple case studies
Holistic	Particularistic
Assumes a dynamic reality	Assumes a stable reality

Each of these different variations toward developing nursing science will determine whether the method fits the phenomena. Generally, nursing phenomena and the human science perspective of nursing are both consistent with the approach that is experiential, qualitative, and contextual. As nursing continues to advance as a human science through theory and research, it must continue to question its old dogmas, transcend its existing paradigms, and refocus its scientific attention to human phenomena that are consistent with the nature of nursing and preservation of humanity. Nursing scholars must seek out alternative methodologies that lead to increased understanding and contribute new human care knowledge that is internally relevant for humans. In doing so, nursing has a great potential to contribute to the growing field of human science and assure itself a place in academic and scientific circles as a scholarly health discipline worthy of advanced studies, independent practice, and epistemic endeavors that serve society.

TABLE 5. MOVEMENT OF VARIOUS DIMENSIONS ON QUANTITATIVE TO QUALITATIVE APPROACHES

Methodology	Quantitative → Qualitative or combination of qualitative-quantitative
Quality Criterion	Rigor (control) → Relevance (meaningfulness) with rigor
Comprehension	Explaining → Understanding
Theory Source	Apriori—deductive—"before the fact" → Grounded field (combination of apriori with openness for new posteriori discoveries from process and data itself)
Knowledge Types	Propositional → Tacit (anticipated and unanticipated happenings)
Instruments	Variety ("layer" between subject and investigator) → (Self is tool part of inquiry) meaning of data becomes measurement
Design	Pre-ordinate (every step described) → Unfolding ("rolling") less structured
Setting	Lab "controlled" setting → Real world, subjective experiences of both subject and investigator

(Adapted from Guba, E.G. & Lincoln, Y.S. Effective Evaluation: Improving the Usefulness of Evaluation Results Through Responsive and Naturalistic Approaches. San Francisco: Jossey-Bass, 1981.)

Healing is not a science, but the intuitive art of wooing nature.

Auden
The Art of Healing

REFERENCES

1. Watson, J. Nursing's scientific quest. Nursing Outlook, 1981, **29**(7), 413–416.
2. McKay, R. Personal communication. Denver, Colo.: University of Colorado Health Sciences Center, 1979.
3. Watson, J. Nursing's scientific quest. Nursing Outlook, 1981, **29**, 413–416.
4. Nightingale, F. Notes on Nursing: What It Is and What It Is Not. New York: Appleton, 1860, 355.
5. Henderson, V. The nature of nursing. American Journal of Nursing, 1964, **64**, 62–68.
6. Goodrich, A. The Social and Ethical Significance of Nursing. New Haven, Conn.: Yale University School of Nursing, 1973, 5. (Originally published 1915.)
7. Giorgi, A. Psychology as a Human Science. New York: Harper & Row, Pub., 1970, 20.
8. Davis, A. The phenomenological approach in nursing research. In N. Chaska

(Ed.), The Nursing Profession: Views Through the Mist. New York: McGraw-Hill, 1978, 186–196.

9. Watson, J. Nursing: The Philosophy and Science of Caring. Boston: Little, Brown, 1979, 205–215.

10. Watson, J. Nursing's scientific quest. Nursing Outlook, 1981, 29, 413–416.

11. Watson, J. Reflections on new methodologies for study of human care. In M. Leininger (Ed.), Qualitative Methodologies in Nursing. New York: Grune & Stratton, 1984 (in press).

12. Winstead-Fry, P. The scientific method and its impact on holistic health. Advances in Nursing Science, 1980, 2, 1–7.

13. Parse, R.R. Man-Living Health: A Theory of Nursing. New York: Wiley, 1981.

14. Webster, G., Jacox, A., & Baldwin, B. Nursing theory and the ghost of the received view. In J. McCloskey & H. Grace (Eds.), Current Issues in Nursing. Scranton, Penn.: Blackwell Scientific Publ., 1981, 26–34.

15. Downs, F.S. It's a great idea—But it won't work. Nursing Research, 1982, 31, 4.

16. Munhall, P.L. Nursing philosophy and nursing research: In apposition or opposition? Nursing Research, 1982, 31, 176–177.

17. Chinn, P. Editorial Advances in Nursing Science, 1983, 5, 2.

18. Donaldson, S. Let us not abandon the humanities. Nursing Outlook, 1983, 31, 40–43.

19. Newman, M. Theory Development in Nursing. Philadelphia: Davis, 1979.

20. Leininger, M. (Ed.). Caring. Thorofare, N.J.: Charles B. Slack, 1981.

21. Watson, J. Reflections on new methodologies for study of human care. In M. Leininger (Ed.), Qualitative Methodologies in Nursing. New York: Grune & Stratton, 1984 (in press).

22. Johnson, R. In Quest of a New Psychology. New York: Human Sciences Press, 1975, 18–19.

23. Koch, S. Psychology: A Study of Science. New York: McGraw-Hill, 1959.

24. Koch, S. Psychology cannot be a coherent science. Psychology Today, 1969, 3, 14, 64, 66.

25. Giorgi, A. Psychology as a Human Science, 40–56.

26. Marton, F. & Svensson, L. Conceptions in research in student learning. Higher Education, 1979, 8, 471–486.

27. Ibid.

28. Cook, T. & Reichardt, C. (Eds.). Qualitative and Quantitative Method on Evaluation Research (Vol. 1). Beverly Hills, Calif.: Sage Publications, 1979, 10.

29. Watson, J. Reflections on new methodologies for study of human care. In M. Leininger (Ed.). Qualitative Methodologies in Nursing.

30. Cook, T. & Reichardt, C. (Eds.). Qualitative and Quantitative Method on Evaluation Research, 9–18.

BIBLIOGRAPHY

Abdellah, F.G. The nature of nursing science. Nursing Research, 1969, 18, 390.

Alexandersson, C. Amadeo Giorgi's empirical phenomenology (Publication No. 3). Swedish Council for Research in Humanities and Social Sciences, Department of Education, University of Goteborg, Sweden, 1981.

Chinn, P. Editorial. Advances in Nursing Science, 1983, 5, 2.

Cook, T. & Reichardt, C. (Eds.). Qualitative and Quantitative Method on Evaluation Research (Vol. 1). Beverly Hills, Calif.: Sage Publications, 1979.

Davis, A. The phenomenological approach in nursing research. In N. Chaska (Ed.), The Nursing Profession: Views Through the Mist. New York: McGraw-Hill, 1978.

Dennis, N. New Methods for Research. Paper presented at Western Australian Institute of Technology, Western Australia, May 1982.

Donaldson, S. Let us not abandon the humanities. Nursing Outlook, 1983, 31, 40-43.

Downs, F.S. It's a great idea—But it won't work. Nursing Research, 1982, 31, 4.

Flaskerud, J.N. & Halloran, E. Areas of agreement in nursing theory development. Advances in Nursing Science, 1980, 3, 1-7.

Gaylin, W. Caring. New York: Knopf, 1976.

Giorgi, A. Psychology as a Human Science. New York: Harper & Row, Pub., 1970.

Guba, E.G. & Lincoln, Y.S. Epistemological and methodological bases of naturalistic inquiry. Educational Communication and Technology Journal, 1982, 30, 233-252.

Hall, L.E. Nursing—What is it? Canadian Nurse, 1964, 60, 150-154.

Henderson, V. The nature of nursing. American Journal of Nursing, 1964, 64, 62-68.

Hyde, A. The Phenomenon of Caring, (Vols. 10-12). American Nursing Foundation, 1975-1977.

Johnson, D. State of art of theory development in nursing. In National League for Nursing (Ed.), Theory Development, What, Why, and How. New York: National League for Nursing, 1978.

Johnson, R. In Quest of a New Psychology. New York: Human Sciences Press, 1975.

King, I. Toward a Theory for Nursing. New York: Wiley, 1971.

Koch, S. Psychology cannot be a coherent science. Psychology Today, 1969, 3, 64, 66.

Koch, S. (Ed.). Psychology: A Study of Science. New York: McGraw-Hill, 1959.

Kreuter, F.R. What is good nursing care? Nursing Outlook, 1957, 5, 302-305.

Leininger, M. (Ed.). Caring. Thorofare, N.J.: Charles B. Slack, 1981.

Leininger, M. Conference on the nature of science in nursing. Introduction: Nature of science in nursing. Nursing Research, 1969, 18, 388-389.

Leininger, M. Foreword. In J. Watson (Ed.), Nursing: The Philosophy and Science of Caring. Boston: Little, Brown, 1979.

Levin, M. Holistic nursing. Nursing Clinics of North America, 6(2), 1971.

Lincoln, Y.S. & Guba, E.G. Understanding and Doing Naturalistic Inquiry. Beverly Hills, Calif.: Sage Publications, 1984.

Marton, F. Describing conceptions of the world around us. Reports from the Institute of Education. University of Goteborg, Goteborg, Sweden, 1978.

Marton, F. & Svensson, L. Conceptions of research in student learning. Higher Education, 1979, 8, 471-486.

Mayerhoff, M. On Caring. New York: Harper & Row, Pub., 1971.

McKay, R. Personal communication. Denver, Colo.: University of Colorado Health: Sciences Center, 1979.

Munhall, P.L. Nursing philosophy and nursing research: In apposition or opposition? Nursing Research, 1982, 31, 176, 177, 181.

Murphy, J. (Ed.). Theoretical Issues in Professional Nursing. New York: Appleton-Century-Crofts, 1971.

Newman, M. Theory Development in Nursing. Philadelphia: Davis, 1979.

Nightingale, F. Notes on Nursing: What It Is and What It Is Not. New York: Appleton, 1860.

Norris, C.M. (Ed.). Proceedings, First Nursing Theory Conference. Kansas City: University of Kansas, 1969.

Oiler, C. The phenomenological approach in nursing research. Nursing Research, 1982, 31, 178, 181.

Omery, A. Phenomenology: A method for nursing research. Advances in Nursing Science, 1982, 5, 49–63.

Parse, R.R. Man-Living Health: A Theory of Nursing. New York: Wiley, 1981.

Paterson, J.D. & Zderak, L.Y. Humanistic Nursing. New York: Wiley, 1976,

Peplau, H. Interpersonal Relations in Nursing. New York: Putnam, 1952.

Pickering, M. Introduction to qualitative research methodology. Paper presented at the meeting of the American Speech-Language and Hearing Association, Detroit, November, 1980.

Rist, RC. On the relations among educational research paradigms: From disdain to detente. Anthropology Educational Quarterly, 1977, 8, 42–49.

Rogers, M. Theoretical Basis of Nursing. Philadelphia: Davis, 1970.

Spicker, S. & Gadow, S. (Eds.), Nursing Images and Ideals. New York: Springer, 1980.

Stevens, B. Nursing Theory. Boston: Little, Brown, 1979.

Valle, R. & King, M. Existential Phenomenological Alternatives for Psychology. New York: Oxford University Press, 1978.

Van Kaam, A.L. Phenomenological analysis: Exemplified by a study of the experience of being really understood. Individual Psychology, 1959, 15, 66–72.

Watson, J. Nursing: The Philosophy and Science of Caring. Boston: Little, Brown, 1979.

Watson, J. Nursing's scientific quest. Nursing Outlook, 1981, 29, (7) 413–416.

Watson, J. Professional identity crisis—Is nursing finally growing up? American Journal of Nursing, 1981, 81, 1488–1490.

Watson, J. Reflections on new methodologies for study of human care. In M. Leininger, (Ed.). Qualitative Methodologies in Nursing. New York: Grune & Stratton, 1984 (in press).

Webster, G., Jacox, A. & Baldwin, B. Nursing theory and the ghost of the received view. In McClosky & Grace (Eds.). Current Issues in Nursing. Boston: Blackwell, 1981.

Winstead-Fry P. The scientific method and its impact on holistic health. Advances in Nursing Science, 1980, 2, 1–7.

Yura, K. & Torres, G. Today's conceptual frameworks within baccalaureate nursing programs. New York: National League of Nursing, 1975, 17–25.

3

HUMAN CARE IN NURSING

Whether humanity is to continue and comprehensively prosper on space-ship Earth depends entirely on the integrity of human individuals and not on political and economic systems. The cosmic question has been asked: Are humans a worthwhile-to-Universe Invention?

R. Buckminster Fuller*

A human science approach to health care is required for nursing practice now and in the future. As the nursing profession advances in the scientific arena as well as in the area of humanistic clinical practice a new model of the nurse is needed. Just as the mind is inseparable from the body, the scholarly activities of nursing should not be divorced from its clinical practice. The new model of nursing for education, research, and practice is that of a scholar–clinician. Likewise, the new wave in health care is an individual approach, directed toward the person that integrates all the parts into a unified and significant whole. Quality nursing and health care today demand a humanistic respect for the functional unity of the human being. The phenomena of health-illness must be approached from a broad conceptual base.

The process of human care for individuals, families, and groups is a major focus for nursing not only because of the dynamic human-to-human transactions, but because of the requirements of knowledge, commitment,

* From the Conference on World Affairs, University of Colorado, April, 1983.

and human values, and because of the personal, social, and moral engagement of the nurse in time and space.

RATIONALE FOR CLARIFYING NURSING

In the 1930s and 1940s America had begun to recognize an epidemic of stress-related illness and disease in our society. Since then stress has been linked to both mental and physical disease, for example, depression, general anxiety, alcoholism, drug addiction, and breakdown in normal relations with friends, family, and colleagues. Unrelieved stress can lead to hypertension, coronary disease, migraine and tension headaches, peptic ulcers, renal disease, asthma, and even cancer. Stress is also related to low productivity, absenteeism, general unhappiness, poor self-worth, failure, helplessness, hospitalization, and premature deaths. As diseases and health–wellness issues increasingly shift from infectious bacteria-linked etiologies to tension-linked etiologies, nursing care becomes an urgent primary and secondary preventive concern. All of this is confounded by a rapidly advancing movement toward health legislation, Diagnostic Related Groups (DRGs), and similar constraints, which must accommodate nursing preventions and interventions with stress-related health–illness concerns.

Along with increased knowledge and awareness of stress-related health problems, is a rapidly growing body of knowledge and skills associated with stress management, coping mechanisms, and stress-reduction strategies. All of these non-medical areas are becoming incorporated into the traditional roles of the nurse.

A human science, metaphysical view of nursing will be adopted in this book as an attempt to develop further and reflect upon the science of nursing, the human care process and human-to-human transactions that become increasingly important in the rapidly growing, complex technological health care systems.

We are all aware that nursing in the United States, Canada, and other industrialized countries has become increasingly technological and bureaucratic. Even community health nursing, which historically had more flexibility and autonomy, has become increasingly managerial and supervisory.

There has been a proliferation of the "curing syndrome" and an adoption of the cure techniques, often without regard to costs. In order to satisfy the increasingly technological and bureaucratic demands of the system, human care at the individual and group level has received less and less emphasis in the system. It is becoming increasingly difficult for nursing to sustain its caring ideology in practice.[1]

Institutions and community health systems alike are organized and administered in a manner that is incongruent with professional human caring. Because of the one-sided perspective of the traditional health care (illness–cure) system, caring values of nurses and nursing have become submerged. Furthermore, the concept of a human care function of the nurse is threatened by the technology, the machines, the high-intensity pace of management, the administrative tasks, and the manipulation of people required to meet the needs of the systems.

Preservation and advancement of human care is a critical issue for nursing today in our increasingly depersonalized society. The mandate for nursing within science as well as within society is a demand for cherishing of the wholeness of human personality. It is thus that I regard nursing as a human science and the human care process in nursing as a significant humanitarian and epistemic act that contributes to the preservation of humanity.

However, some of the critical epistemological, scientific, ethical, esthetic, methodological, and even metaphysical questions have yet to be explored by nursing. At the same time these questions that remain to be explored are related to the nature of the subject matter of human nurse and human patient or person and how they are defined or codefined.

Furthermore, what are the conditions that facilitate and sustain a person as an end in and of him- or herself and not as a means to some scientific, medical, nursing, or hospital end? What are the conditions that facilitate or sustain human care and caring in instances of threatened humanity? What are the conditions that sustain a person and human caring in instances of threatened biological–organic states? What are the nursing conditions that collectively facilitate and sustain the preservation of humanity in instances of threatened humanity?*

Nursing as a human science and human care is always threatened and fragile. Because human care and caring requires a personal, social, moral, and spiritual engagement of the nurse and a commitment to oneself and other humans, nursing offers the promise of human preservation in society.

What I am as a care provider and a caring person and nurse now is, and must be, connected with what I will be for another in the future. The now of human care and caring shapes the future and the ontology of caring in time and space.

Human caring in nursing, therefore, is not just an emotion, concern, attitude, or benevolent desire. Caring is the moral ideal of nursing whereby the end is protection, enhancement, and preservation of human dignity.[2] Human caring involves values, a will and a commitment to care, knowledge, caring actions, and consequences. All of human caring is related to intersubjective human responses to health–illness conditions; a knowledge of health–illness, environmental–personal interactions; a knowledge of the nurse caring process; self-knowledge, knowledge of one's power and transaction limitations.

---

* These ideas were influenced by Professor Gary Stahl, Department of Philosophy, University of Colorado, Boulder.

To quote Mayerhoff:

> We sometimes speak as if caring did not require knowledge, as if caring
> for someone, for example, were simply a matter of good intentions or warm
> regard. . . . To care for someone, I must know many things. I must know,
> for example, who the other is, what his powers and limitations are, what
> his needs are, and what is conducive to his growth; I must know how to
> respond to his needs and what my own powers and limitations are. Such
> knowledge is both general and specific.[3]

As such human care is an epistemic endeavor that defines both nurse and
person and a level of space and time, requires serious study, reflection,
action, and a search for new knowledge and new insights that will help
to discover new meanings and understanding of the person and human
care process during health–illness experiences.

Such a search for a new knowledge and understandings of human care
will then govern some of the epistemological, ethical, intuitive, esthetic,
scientific, and methodological conditions for developing nursing as a
human science. Moreover, such epistemic endeavors in nursing will help
to shed light on both the nurse and the patient or person as valuable ends
in and of themselves, who are coactive and codeterminant partners in the
human–care process. It is the interdependent, intersubjective human
process, therefore, that can shape conditions necessary to sustain a person and caring in instances where humanity is threatened.

REFERENCES

1. Ray, M. A philosophical analysis of caring within nursing. In M. Leininger,
   (Ed.), Caring: An Essential Human Need. Thorofare, N.J.: Charles B. Slack,
   1981, 25–36.
2. Gadow, S. Existential advocacy as a form of caring: Technology, truth, and
   touch. Paper presented to the Research Seminar Series: The Development of
   Nursing as a Human Science. School of Nursing, University of Colorado Health
   Sciences Center, Denver, March, 1984.
3. Mayerhoff, M. On Caring. New York: Harper & Row, Pub., 1971, 13.

BIBLIOGRAPHY

Gadow, S. Existential advocacy as a form of caring: Technology, truth, and touch.
   Paper presented to the Research Seminar Series: The Development of Nursing
   as a Human Science. School of Nursing, University of Colorado Health Sciences
   Center, Denver, March, 1984.
Mayerhoff, M. On Caring. New York: Harper & Row, Pub., 1971.
Ray, M. A philosophical analysis of caring within nursing. In M. Leininger, (Ed.),
   Caring: An Essential Human Need. Thorofare, N.J.: Charles B. Slack, 1981.

4

NATURE OF HUMAN CARE
AND CARING VALUES
IN NURSING

*The only true standard of greatness of any civilization is our sense of social
and moral responsibility in translating material wealth to human values
and achieving our full potential as a caring society.*

*The Right Honorable Norman Kirk**

This chapter discusses the values underlying human care and human
science in nursing. The basic assumptions related to human care values
are put forth, along with some acknowledgment of the dual nature of the
relationship between caring and noncaring.

A recognition and acknowledgment of the value of human care in nursing comes before and presupposes actual caring. A nurse may perform
actions toward a patient out of a sense of duty or moral obligation, and
would be an ethical nurse. Yet it may be false to say he or she cared about
the patient. The value of human care and caring involves a higher sense
of spirit of self. Caring calls for a philosophy of moral commitment toward
protecting human dignity and preserving humanity. According to Gaut
the general family of meanings related to the notion of caring include: individual attention to and concern for; individual responsibility for or providing for at some level; individual regard, fondness or attachment.[1]

The ideal and value of caring is clearly not just a *thing* out there, but
is a starting point, a stance, an attitude, which has to become a will, an

---

* From a speech by the Right Honorable Norman Kirk, the former Prime Minister of New
Zealand.

intention, a commitment, and a conscious judgment that manifests itself in concrete *acts*. Human care, as a moral ideal, also transcends the act and goes beyond the specific act of an individual nurse and produces collective acts of the nursing profession that have important consequences for human civilization.

The essence of the value of human care and caring may be futile unless it contributes to a philosophy of action. Gaut goes further and says the action must be judged solely on the welfare of the person being cared for.[2]

The actual concrete action of caring can transcend the value (and pass it on). Embedded in this idea is the notion that caring values and actions can be contagious, at an individual and systemic level, if sufficient conditions are met. The value of caring is grounded in the self-transcending creative nurse. Gaut indicates that the necessary and sufficient conditions for caring include:

1. Awareness and knowledge about one's need for care
2. An intention to act, and actions based on knowledge
3. A positive change as result of caring, judged solely on basis of welfare of others.[3]

I would add that there must be an underlying value and moral commitment to care and a will to care.

In order for nursing to be truly responsive to the needs of society and make contributions that are consistent with its roots and early origins, both nursing education and the health care delivery system must be based on human values and concern for the welfare of others. Caring outcomes in practice, research, and theory depend on the teaching of a caring ideology.

As the human threats from biotechnology, scientific engineering, fragmented treatment, bureaucracy, and depersonalization increase and spread in our health care delivery system, so must we increase and spread the human care philosophy, knowledge and practices in our systems.

Nursing is the profession that has an ethical and social responsibility to both individuals and society to be the caretaker of care and the vanguard of society's human care needs now and in the future.

Assumptions Related to Human Care Values in Nursing:

1. Care and love are the most universal, the most tremendous, and the most mysterious of cosmic forces: they comprise the primal and universal psychic energy.[4]
2. Often these needs are overlooked; or we know people need each other in loving and caring ways, but often we do not behave well toward each other. If our humanness is to survive, however, we need to become more caring and loving to nourish our humanity and evolve as a civilization and live together.[5]

3. Since nursing is a caring profession, its ability to sustain its caring ideal and ideology in practice will affect the human development of civilization and determine nursing's contribution to society.
4. As a beginning we have to impose our own will to care and love upon our own behavior and not on others. We have to treat ourself with gentleness and dignity before we can respect and care for others with gentleness and dignity.[6]
5. Nursing has always held a human-care and caring stance in regard to people with health–illness concerns.
6. Caring is the essence of nursing and the most central and unifying focus for nursing practice.[7]
7. Human care, at the individual and group level, has received less and less emphasis in the health care delivery system.
8. Caring values of nurses and nursing have been submerged. Nursing and society are, therefore, in a critical situation today in sustaining human care ideals and a caring ideology in practice. The human care role is threatened by increased medical technology, bureaucratic–managerial institutional constraints in a nuclear age society. At the same time there has been a proliferation of curing and radical treatment cure techniques often without regard to costs.
9. Preservation and advancement of human care as both an epistemic and clinical endeavor is a significant issue for nursing today and in the future.
10. Human care can be effectively demonstrated and practiced only interpersonally. The intersubjective human process keeps alive a common sense of humanity; it teaches us how to be human by identifying ourselves with others, whereby the humanity of one is reflected in the other.
11. Nursing's social, moral, and scientific contributions to humankind and society lie in its commitment to human care ideals in theory, practice and research.

CARING AND NONCARING

It is important to notice the dual nature of the relation between caring and noncaring. What we call caring on one occasion must be the same as what we call caring on another occasion. That which nursing sets before itself is an ideal that it is trying to reach. Caring must be the same thing that one achieves at one moment when one is caring or fails to achieve when one fails to be caring.

We can also approach caring by the method of contrast. A distinction can be found in society (and in nursing) in some form, between those persons (nurses) who are caring and those who are uncaring. The most abstract characteristics of a caring person is that he or she is somehow responsive to a person as a unique individual, perceives the other's feelings, and sets apart one person from another from the ordinary. The uncaring person is by contrast insensitive to another person as a unique individual, nonperceptive of the other's feelings and does not necessarily distinguish one person from another in any significant way.

Empirical research on caring in nursing has substantiated the above view that caring indeed connotes a personal response. The early findings of Watson, Burckhardt, Brown, et al. in 1979, derived caring categories from empirical data which revealed the following process: "treating the individual as a person," "concern and empathy," "personalized characteristics of the nurse," "communication process," and "extra effort."[8]

Moreover, Watson's 1983 cross-cultural data on caring in nursing supports some of the above findings.[9] A study of human caring among Australian Anglo-Saxons and Aborigines and Chinese in Taiwan revealed strong and consistent results linking caring to personal responses.

The cross-cultural data included categories such as "nurse presence" (including touch, and physical "felt presence" from nurse across time; for example, sharing an experience across time "nurse feelings exchanged," exchange of love, sharing sorrow, pain; letting a person feel; and a category related to caring nurses "giving time and taking time"—this category included follow-up checks, presence, visitation). All of these findings are consistent with an individual approach, and conscious acts that convey a will and intention to care, along with specific actions.

Both theoretically and empirically the concept of caring is not merely characterized by certain categories or classes of nursing actions, but as ideals, which persons desiring care and persons (nurses) doing those actions hold before them. Caring actions themselves and the various ways caring is revealed are not so much instances of caring per se as approximations of caring.

Perceptible characteristics of caring that are revealed in actions and attitudes are some particulars, but are not a pure form of caring. Nevertheless, such views and data on caring provide us with ways to contrast caring with noncaring to get a better understanding of the phenomena.

WATSON'S VALUE SYSTEM

The value system set forth here regarding a theory of human care consists of values associated with deep respect for the wonders and mysteries of life; acknowledgment of a spiritual dimension to life and internal power of the human care process; growth and change. Human care requires high regard and reverence for a person and human life, nonpaternalistic values that are related to human autonomy, and freedom of choice. There is a high value on the subjective–internal world of the experiencing person and how the person (both patient and nurse) is perceiving and experiencing health–illness conditions. An emphasis is placed upon helping a person gain more self-knowledge, self-control, and readiness for self-healing, regardless of the external health condition. The nurse is viewed as a co-participant in the human care process. Therefore, a high value is placed on the relationship between the nurse and the person.

This value system is blended with Watson's ten carative factors,[10] such as humanistic altruism; sensitivity to oneself and others; along with love for and trust of life and other humans.

Underlining the value system is a call for a revaluing of humans and human caring in theory, practice and science—thus a rationale for developing nursing as a human science wherein the person is the starting point.

Such a perspective leads to some metaphysical considerations that are necessary to discuss before moving forward.

REFERENCES

1. Gaut, D. Development of a theoretically adequate description of caring. Western Journal of Nursing Research, 1983, 5(4), 315.
2. Ibid., 313–324.
3. Ibid.
4. De Chardin T. On Love. New York: Harper & Row, Pub., 1967, 7–8.
5. Ibid.
6. Ibid.
7. Leininger, M. (Ed.). Caring: An Essential Human Need. Thorofare, N. J.: Charles B. Slack, 1981.
8. Watson, J., Burckhardt, C., Brown, I., et al. A model of caring. In the American Nurses' Association, Clinical and Scientific Sessions. Kansas City, Mo.: American Nurses' Association, 1979, 32–44.
9. Watson, J. Caring and loss-grieving experiences, new knowledge for nursing practice. Research presented at the American Nurses' Association's Clinical and Scientific Sessions. Denver, Colo., November 1983.
10. Watson, J. Nursing: The Philosophy and Science of Caring. Boston: Little, Brown, 1979, 9–10.

BIBLIOGRAPHY

De Chardin, T. On Love. New York: Harper & Row, Pub., 1967.
Gaut, D. Development of a theoretically adequate description of caring. Western Journal of Nursing Research, 1983, 5(4).
Leininger, M. (Ed.). Caring: An Essential Human Need. Thorofare, N. J.: Charles B. Slack, 1981.

Watson, J., Burckhardt, C., Brown, L., et al. A model of caring. In the American Nurses' Association, Clinical and Scientific Sessions. Kansas City, Mo.: American Nurses' Association, 1979.

Watson, J. Caring and loss-grieving experiences, new knowledge for nursing practice. Research presented at the American Nurses' Association Clinical and Scientific Sessions. Denver, Colo., November 1983.

Watson, J. Nursing: The Philosophy and Science of Caring. Boston: Little, Brown, 1979.

5

NURSING AND METAPHYSICS

Still another such need, strangely, is the need for metaphysics her-self, ... What am I? What is death—and more puzzling still, what is birth? A beginning? An ending? ... does it matter?

Richard Taylor
Metaphysics*

The previous chapters have attempted to set forth a view of nursing that is consistent with nursing's tradition of human caring, rather than the tradition of medicine. In advancing such a view there is a call for a revaluation of humans and caring. The alternative world view of nursing that is being suggested will place nursing within a metaphysical context and establish nursing as a human-to-human care process with spiritual dimensions, rather than a set of behaviors that conform to the traditional science/medical model.

Western science, psychology, and even nursing have dealt very poorly with the spiritual side of human nature; it is either ignored or labeled pathological, too religious, too abstract, too extreme, or controversial. Yet much of the agony of our time stems from a spiritual vacuum. Western culture and our study of people and nursing has ruled out spiritual nature, but the cost is great.[1]

A metaphysical context becomes important not only to me in my writing, but my context gives the reader a better perspective of my values

---

* Taylor, R. Metaphysics (2nd Ed.). Englewood Cliffs, N.J.: Prentice-Hall, 1974.

and beliefs and the opportunity to assess how these mesh with my ideas and with one's own ideas. It also becomes useful for any nurse, be he or she a scholar, teacher, researcher, or practitioner of nursing to step back and examine nursing as a professional, social, and scientific endeavor that exists as a service to humankind. As such, it is necessary to examine and reflect upon what nursing is, does, stands for, and could or should contribute to society.

The questions beyond the starting point of my ideas are, What should nursing be about? What change of values, goals, and visions are required for nursing to actualize its true sense of direction, its true social and scientific contribution? What lens change is necesary? What point of view needs examining? Does any of this require a new starting point, or do we need greater lens power "to see" or "to be"? All of this, of course, fits within a broader context of society, environment, culture, politics, time, and space.

My nursing views are of the ideal, what may be, rather than what is, but they also acknowledge that what exists as the essence and power of nursing is underdeveloped and often overlooked.

ROLE OF METAPHYSICS IN WESTERN SCIENCE

Most people today are aware that with the rise of natural sciences throughout the history and philosophy of science, the positivistic-reductionist approach dominated Western science and medicine and affected nursing. Progressive thinkers who incorporated metaphysical beliefs into their ideas were often disregarded or rejected. Because of the positivist tradition of Western science and the advancement of medical science, Eastern thought that incorporated the spiritual aspect of humanness has had little impact on nursing. However, Eastern ideas and philosophies are very common in the writings of different 19th and early 20th century poets and authors, including such transcendentalists as Ralph Waldo Emerson, Henry David Thoreau, Walt Whitman, and to some extent William James. Even Florence Nightingale offered a metaphysical orientation when she emphasized that nature restored and preserved health. In essence the nurse was to be good and loving; "go your way straight to God's work in simplicity and singleness of heart."[2] Other early historians of nursing posited that the principal condition in human survival is human care.[3]

Among modern theorists, C. G. Jung certainly had a strong orientation toward Eastern thought and religions. Jung talked about people facing their souls.[4]

So while the 19th century writers and poets glimpsed "cosmic consciousness" (a phrase Walt Whitman borrowed from Vedantic philosophy of India)[5] only more recently have these ideas been of interest to Western thinkers and scientists and nurses.

Medical science has moved from an integrated approach in the early stages—the physician as healer and priest era (where the mind, body, and soul were united for care and cure)—to the period of scientism where they split apart and different specialists, different health care providers, or different technologies or medical treatments were applied to the different components of the person. In the current period the person is split further and further apart and the soul is either replaced with narcissism of self or denied altogether. The human soul is further destroyed with a depersonalized, manmade environment, advanced technology, and robot treatment for cure, delivered by strangers in a strange environment.

The developmental trends of humankind throughout history indicate that an awareness of psychological or mental processes comes about more slowly than an awareness of physical concerns. Just as history seems to repeat itself, movements tend to be evolutionary, rather than revolutionary, cyclical and spiral. Take, for example, the idea of "self," which is a relatively recent term in psychology. Although some late 18th century and early 20th century literary works were among the first to develop the concept of self (Jane Austen, Henry James, William James), the notion of self was picked up later by the more scientific world of psychology and sociology. Perhaps Allport and Maslow in the 1950s and 1960s were the first psychologists really to begin to focus on the notion of self. This was followed by the human potential movement in psychology and related fields exemplified by Carl Rogers, and Fritz Perls, the Esalen movement, and so on; all were attempts to uncover, discover, recover, and restore one's sense of self. During the evolution of scientific–philosophical advancements, coupled with the changing world views of what it means to have a self, and to be human, nursing has possessed the primary responsibility of caring for people in health and illness. The degree to which the nurse cared for patients, the focus of what was important for caring changed and continued to change. However, the aspect of lasting significance is that nurses cared for and about others.

The thesis of this work is that caring as an intersubjective human process is the moral ideal of nursing. Nursing, therefore, has an important social role in society in the enhancement of dignity and the preservation of humanity. Indeed nursing's role in society is based on human caring; its social contribution lies in its moral commitment to human care. It also has an important humanistic and scientific contribution to make in the field of human sciences and health sciences in pursuing human care as a serious epistemic endeavor. This scientific awareness has been long in coming and slow to be recognized, and still has far to go. Nevertheless, human care theory and knowledge can now be viewed as a significant scientific pursuit. Moreover, there is more social, artistic, literary, and scientific freedom and permission to attend to moral and metaphysical matters today than in the 1960s, with its scientism rise. There is now a recognition of an inner self, inner resources, acknowledgment of spiritual self, or the need for integration of the mind, body and soul. Our concept of human development need not stop with ideas of self and self-actualization, but can allow for spiritual awakening and pursuit of harmony among the mind, body, and soul.

It is ironic that in a time of such tangible, factual, scientific, and technological advancement in medical science, we have to turn to some sense of mysterious, intangible, philosophical, and metaphysical, sometimes even mystical, worlds of humans to solve some of the sickness in society, the suffering associated with disharmony with the mind, body, and soul.

As a result of the historical movements, nursing is now at a point where it can also consider some metaphysical and moral ideals as guides for its own efforts. Nursing science can benefit from a metaphysical approach that revalues the higher spiritual sense of being human, links that with human care as it advances itself as a human science for the 21st century. Nursing can in turn help justify its human concern to build a more fully informed understanding of the spiritual realms and where that leads us as humans and scientists.

The nursing theory I am developing attempts to make explicit my metaphysical position regarding mind, body, and soul and tries to acknowledge how my beginning position on this issue directs my concept of nursing and caring processes.

To illustrate the complexities of metaphysical issues that confront nursing, whether they are acknowledged or not, I am including Chapter 1, The Need for Metaphysics, from Richard Taylor's *Metaphysics*.[6] This chapter can be a beginning context for nurses to consider their own views on the complex matter of being and knowing, and to learn how it is impossible to escape metaphysics, regardless of how hard one tries or how many substitutes are created. The chapter, which follows, helps one to examine the need for metaphysics which can guide one's response to others and affect nursing theories, practice, and research.

THE NEED FOR METAPHYSICS*
Richard Taylor

There are many things one can do without. Among them are even things foolish persons devote their life's energy to winning. One can do without wealth, for example, and be no less happy. One can do without position, status, or power over others, and be no less happy, certainly no less human.

---

*From Taylor, R. Metaphysics (2nd Ed.). Englewood Cliffs, N.J.: Prentice-Hall, 1974, pp. 5-9, with permission.

Very likely, without these one will be more human, more the kind of being nature or God, whatever gave him being in the first place, intended. Indeed such things as these—possessions, power, notoriety, which mean so much to the unreflective—appear on examination to be no more than desperate attempts to give meaning to a life that is without meaning. They reflect the vain notion that one's worth can be protected, even enhanced, by enough accumulation, if not of gold, then of its modern equivalents. When this fails, the pursuit of such things as often as not becomes little more than the response to the need to have something to do. Few people are able to sit still, much less to sit still and think; and when enforced idleness threatens, most people begin to plan distant places to go to, purchases to be made, or pictures to take in far off lands—in short, something to do. Just the going and coming will keep them busy for a while, get them through that much of life, and take their minds off things by presenting a variety and novelty to their sense organs. Perhaps man is, as the ancients declared, a rational animal; but if this is so, it is only in the sense that he is uniquely capable of reason, contemplation, and thought, not that he spends much of his life at it. We still share with the rest of animate creatures restless needs and cravings that drive us to movement perpetually. Aristotle's dictum that life is motion surely applies to our own lives. It is what we share with other animals that is most apparent, not the elusive qualities that set us apart. The same philosopher associated reason and thought and contemplation with the gods. He did not first look to mortals for the expression of intelligence—except, interestingly, to those few of its specimens gifted with the love for philosophy and metaphysics, and whose happiness he therefore compared to that of the gods.

The Love of Man and of Nature

There are, however, some things one cannot do without, at least not without deep suffering and the diminishing of one's nature. Among these is the love and approbation of at least a few of one's fellows. Lacking these, one seeks the semblance of them in the form of feigned affection, pretended deference, awe, and sometimes fear. It is astonishing that these counterfeits will so often do, will even seem to give significance to people's lives. Yet we see on every side that this is in fact so. The explanation is of course not in the worth of these things themselves, but in the depth of the need people vainly seek to satisfy through the means of such things.

Another need that cannot be destroyed or left unmet without great damage, of which metaphysicians have often been acutely aware, is the love for nature and the feeling of our place within it. Without this we become machines, grinding out our days and hours to that merciful end when death imposes the peace we have never been able to find for ourselves. A child easily thinks of himself as something apart, a virtual center of reality about which the whole of nature turns, to whose wants everything ministers. One who loves nature rises above this paltry conception of his own being and becomes sensitive to his identity with the whole of reality, which is without beginning or end. This partially explains the difficulty

many persons have in fathoming metaphysics. It is not that it is so difficult, but that it is approached from the wrong point of view—from a child-
hood mentality, from the standpoint of one who finds himself always at
the center of the stage, all else being a vast thing without spirit or soul.
It is hardly the frame of mind in which to understand a Plato, a Buddha,
or a Spinoza.

Metaphysics and Wisdom

Still another such need, strangely, is the need for metaphysics herself.
We cannot live as fully rational men without her. This does not mean that
metaphysics promises the usual rewards that a scientific knowledge of
the world so stingily withholds. She does not promise freedom, God, immortality, or anything of the sort. She offers neither a rational hope nor
the knowledge of these. Metaphysics in fact promises no knowledge of
anything. If knowledge itself is what one seeks, he should be grateful for
empirical science, for he will never find it in metaphysics.

Then what is her reward? What does metaphysics offer that is in her
power alone to give? What, that this boundless world cannot give even
to the richest and most powerful—that she seems, in fact, to withhold
from these more resolutely than from the poor and the humble? Her reward is wisdom. Not boundless wisdom, not invincible truth, which must
be left to the gods, not a great understanding of the cosmos or of man,
but wisdom, just the same; and it is as precious as it is rare.

What, then, is so good about it? What is wisdom worth if it does not
fulfill our deep cravings, such as the craving for freedom, for gods to worship, for a bit more of life than material nature seems to promise? What
makes it worth seeking at all?

The first reward of such wisdom is, negatively, that it saves one from
the numberless substitutes that are constantly invented and tirelessly
peddled to the simple-minded, usually with stunning success, because
there is never any dearth of customers. It saves us from these glittering
gems and baubles, promises and dogmas and creeds that are worth no
more than the stones under one's feet. Fools grasp, at the slightest solicita-
tion, for any specious substitute that offers a hope for the fulfillment of
their desires, the products of brains conditioned by greed and competi-
tion, no matter how stupid, sick, or destructive these may be. Many persons, in response to the deep need to be loved, of which we have spoken,
have felt themselves transformed by a mere utterance; such as, for example, "Jesus loves you!", an assurance that is cheaply and insincerely
flung at them by an ambitious evangelist. The instant conviction that
such bandishments sometimes produce is uncritically taken to be a sign
of their certain truth, when in fact they signify nothing more than a need
which demands somehow to be met, by whatever means. Again, many
persons can banish at will, even before it is really felt, the dread and the
objective certainty of their own inevitable destruction. For this comfort
they need nothing more than the mere reminder of some promise expressed
in a text of ancient authority, or some holy book, or even the simple
declamation of a clever and manipulative preacher. In this way does the

religion of faith, perverting everything and turning the world upside down, serve as the cheap metaphysics, not of the poor, but of those impoverished in spirit and wanting in wisdom, some of whom bask in a blaze of worldly glory. Such religion, substituting empty utterance for thought, is not the religion of the metaphysical mind or of those who love God and nature first and themselves as a reflection of this.

Where religion can make no headway, in the mind of the skeptic, ideology can sometimes offer some sort of satisfaction to much the same need. Thus many persons spend their lives in a sandcastle, a daydream, in which every answer to every metaphysical question decorates its many mansions. The whole thing is the creation of their brains, or worse, of their needs—it is an empty dream, for nothing has been created except illusions. Such dreams are not metaphysics, but the substitute for metaphysics. They illustrate again how one can live without metaphysics only if some substitute, however specious, is supplied, and this is testimony to the deep need for her.

What am I? What is this world, and why is it such? Why is it not like the moon—bleak, barren, hostile, meaningless? How can such a thing as this be? What is this brain; does it think? And this craving or will, whence does it arise? Is it free? Does it perish with me, or not? Is it perhaps everlasting? What is death—and more puzzling still, what is birth? A beginning? An ending? And life—is it a clockwork? Does the world offer no alternatives? And if so, does it matter? What can one think about the gods, if anything at all? Are there any? Or is nature herself her own creator, and the creator of me; both cradle and tomb, both holy and mundane, both heaven and hell?

The answers to such things are not known. They never will be. It is pointless to seek the answers in the human brain, in science, or in the pages of philosophy and metaphysics. But they will be sought, just the same, by everyone who has a brain, by the stupid as well as the learned, by the child, the man, by whoever can look at the world with wonder. False and contrived answers will always abound. There will always be those who declare that they know the answers to these things, that they "found" these answers in some religious experience, in some esoteric book of "divine" authorship, or in something occult. They do not find them; they find nothing at all except the evaporation of their need to go on asking questions, and of their fears of what the answers to those question could turn out to be. They find, in other words, a comfort born of ignorance.

So the need of metaphysics endures. No one will shake it off. Metaphysics will be shunned by most people, always, because her path is not easy and no certain treasures lie at the end. Her poor cousins will be chosen in preference, because they offer everything at no cost—a god to worship who has set us apart from the rest of creation and guarantees each an individual immortality, and a will that is free to create a destiny.

People will always choose substitutes for metaphysics. Because of the indestructible need for her, they will accept anything, however tawdry, however absurd, as a surrogate. Yet it is only metaphysics that, while preserving one in the deepest ignorance, while delivering up not the smallest grain of knowledge of anything, will nevertheless give that which alone is worth holding to, repudiating whatever promises something better. For metaphysics promises wisdom, a wisdom sometimes inseparable from ignorance, but whose glow is nevertheless genuine, from itself, not borrowed, and not merely the reflection of our bright and selfish hopes.[6]

Because of the nature of this text, with its emphasis on humans and human care in nursing, it is inevitable that there are metaphysical overtones. When a discipline's primary subject matter is tied to humans, life, death, and such abstract notions as health and illness and human care processes, it is impossible to limit ideas strictly to empirical sciences, and the physical–materialistic view of life.

One also seeks some greater, deeper reflection, explanations, meanings, and sense of wisdom that goes beyond the knowledge, facts, and external events per se. As such, it is necessary to be forthright about one's metaphysical beliefs; human life, then, becomes the foundation of my ideas about nursing as a deeply human activity.

REFERENCES

1. Tart, C. (Ed.). Transpersonal Psychologies. New York: Harper & Row, Pub., 1976.
2. Nightingale, F. Notes on Nursing: What It Is and What It Is Not. New York: Appleton, 1860, 135–136.
3. Dock, L. & Stewart, I.M. A Short History of Nursing (Vol. 1). New York: Putnam, 1920.
4. Jung, C.G. Psychology and alchemy. In H. Read, M. Fordham, & G. Adler (Eds.), Collected Works of C.G. Jung (Vol. 12). Princeton, N.J.: Princeton University Press, 1968, 99–101.
5. Hall, C.S., & Lindsay, G. Theories of Personality (3rd Ed.). New York: Wiley, 1978, 351.
6. Taylor, R. Metaphysics (2nd Ed.). Englewood Cliffs, N.J.: Prentice-Hall, 1974, 5–9.

BIBLIOGRAPHY

Dock, L. & Stewart, I.M. A Short History of Nursing (Vol. 1). New York: Putnam, 1920.
Hall, C.S. & Lindsay, G. Theories of Personality (3rd Ed.). New York: Wiley, 1978.
Jung, C.G. Psychology and alchemy. In H. Read, M. Fordham, & G. Adler (Eds.), The Collected Works of C.G. Jung (Vol. 12). Princeton, N.J.: Princeton University Press, 1968.
Nightingale, F. Notes on Nursing: What It Is and What It Is Not. New York: Appleton, 1860.
Tart, C. (Ed.). Transpersonal Psychologies. New York: Harper & Row, Pub., 1976.
Taylor, R. Metaphysics (2nd Ed.). Englewood Cliffs, N.J.: Prentice-Hall, 1974.

6

NATURE OF HUMAN LIFE
AS SUBJECT MATTER
OF NURSING

The most beautiful thing we can experience is the mysterious. It is the source of all true art and science.

Anonymous

BASIC BELIEFS

My theory of human care begins with my view of personhood and human existence; that in itself becomes metaphysical.

What is essential in human existence is that the human has transcended nature—yet remains a part of it. The human can go forward, through the use of the mind, to higher levels of consciousness, by finding meaning and harmony in existence.

My conception of life and personhood is tied to notions that one's soul possesses a body that is not confined by objective space and time. The lived world of the experiencing person is not distinguished by external and internal notions of time and space, but shapes its own time and space, which is unconstrained by linearity. Notions of personhood, then, transcend the here and now, and one has the capacity to coexist with past, present, future, all at once. As a result of this view, there is a great deal of regard, respect, and awe given to the concept of a human soul (spirit, or higher sense of self) that is greater than the physical, mental, and emotional existence of a person at any given point in time. The individual spirit of a person or of collective humanity may continue to exist throughout time, keeping alive a higher sense of humankind. Although a body may die, be murdered, kill itself, be diseased, infirmed, and so on, the soul or spirit continues to live on. However, the soul can be underdeveloped, dormant, and in need of reawakening.

According to Jung,

> People will do anything, no matter how absurd, in order to avoid facing their own souls. They will practice yoga and all its exercises, observe a strict regime of diet, learn theosophy by heart, or mechanically repeat mystic texts from the literature of the whole world—all because they cannot get on with themselves and have not the slightest faith that anything useful could ever come out of their own souls.[1]

The belief that a person possesses a soul is to be regarded with the deepest respect, dignity, mystery, and awe because of the continuing, yet unknown, journey throughout time and space, infinite and external. The soul, then, exists for something larger, greater and more powerful than physical life as we know it and could know it for time past, time present, and time future.

The concept of the soul, as used here refers to the *geist*, spirit, inner self, or essence of the person, which is tied to a greater sense of self-awareness, a higher degree of consciousness, an inner strength, and a power that can expand human capacities and allow a person to transcend his or her usual self. The higher sense of consciousness and valuing of inner self can cultivate a fuller access to the intuitive and even sometimes allow uncanny, mystical, or miraculous experiences, modes of thought, feelings, and actions that we have all experienced at some points in our life, but from which our rational, scientific cultures bar us. The terms soul, inner-self, spiritual self, and geist all refer to the same phenomenon and tend to be used interchangeably.

One's ability to transcend space and time occurs in a similar manner through one's mind, imagination, and emotions. Our bodies may be physically present in a given location or situation, but our minds and related feelings may be located elsewhere.

Each of the assumptions underlying the view of human life is that each of us is a magnificent spiritual being who has often been undernourished and reduced to a physical, materialistic being. We know both rationally and intuitively, however, that a person's human predicament may not be related to the external, physical world as much as to the person's inner world as lived and experienced. Awareness of oneself as a spiritual being opens up infinite possibilities.

Poets, sages, and philosophers throughout time have referred to the spiritual side of life and living, and have advocated the self-knowledge, self-reverence, and self-control that comes from the inner, spiritual self. The notion of a spiritual self and inner power requires a different starting point for how we view people, existence, life, and the world. The idea of transcendence is fairly alien to the Western world with its mind–body schism. Yet ancient civilizations, philosophers, and poets have long believed, practiced, and written about the transcendence of self, higher consciousness, over-soul, spiritual experience, mystical experiences, and so on.

The idea of transcendence represents options for true human growth, opportunities to become more fully human. The views inherent in these ideas allow one to turn inward and regard oneself and others with reverence and dignity, as spiritual beings, capable of contributing to the spiritual evolution of self and civilization.

LIFE

Human life in this instance, then, is defined as (spiritual–mental–physical) being-in-the-world, which is continuous in time and space. Only to the extent that a person has fulfilled the concrete meaning of human existence will the self be fulfilled; the meaning that a being has to fulfill is something beyond the self, it is never just self.

The approach herein incorporates scientific views and research with a deep philosophy of the goodness of humankind with a sense of esthetics and metaphysics.

Because of these values and beliefs about life and personhood, it follows that access to the higher sense of self comes more readily through the human emotions, the mind, and the subjective inner world of the experiencing person.

In developing a theory of nursing, it is helpful to clarify one's values and views of human life because those underlying values and beliefs give direction and meaning to nursing, the human care process, and other components of the theory.

The human care process between a nurse and another individual is a special, delicate gift to be cherished. The human care transactions provide a coming together and establishment of contact between persons; one's mind–body–soul engages with another's mind–body–soul in a lived moment. The shared moment of the present has the potential to transcend time and space and the physical, concrete world as we generally view it in the traditional nurse–patient relationship.

Each person (both the nurse and the patient) brings to the present moment his or her own unique causal past.* Each experiential moment of "now" becomes incorporated into one's causal past and helps to direct

---

* This term derives from Whitehead and involves collective but unique past experiences and events that each person brings to the present moment. Each person's causal past and presentational immediacy has the potential to influence the future.

one's future. All three phases of time (past, present, future) can be, and usually are, operating in the inner and lived world of the experiencing person. A person's inner world can transcend time past, present, future, through introspection, creative imagination, meditation, visualization, and the projection of the self in a series of experiences, as well as in sleep, dreaming, and fantasizing, including unconscious and possibly supraconscious processes not yet known or fully explored.

ILLNESS

Illness is not necessarily disease. Illness is subjective turmoil or disharmony within a person's inner self or soul at some level or disharmony within the spheres of the person, for example, in the mind, body, and soul, either consciously or unconsciously. In a situation where one's "I" is separated from one's "me", the self is separated from the self. Illness connotes a felt incongruence within the person such as an incongruence between the self as perceived and the self as experienced.

A troubled inner soul can lead to illness, and illness can produce disease. Specific experiences, for example, developmental conflicts, inner suffering, guilt, self-blame, despair, loss, and grief, and general and specific stress can lead to illness and result in disease. Unknowns can also lead to illness; the unknown can only be known by experience and may require inner searching to find. Disease processes can also result from genetic, constitutional vulnerabilities and manifest themselves when disharmony is present. Disease itself in turn creates more disharmony.

HEALTH

Health refers to unity and harmony within the mind, body, and soul. Health is also associated with the degree of congruence between the self as perceived and the self as experienced.

Such a view of health focuses on the entire nature of the individual in his or her physical, social, esthetic, and moral realms—instead of just certain aspects of human behavior and physiology. Such a view is refered to as an *euda imonistic* model of health.[2]

In summary:

I = Me—Health (harmony, with world and open to increased diversity)

I ≠ Me—Illness (in varying degrees and person less open to increased diversity) When I ≠ me for continuous periods of time, disease may be present.

GOAL

The goal of nursing proposed is to help persons gain a higher degree of harmony within the mind, body, and soul which generates self-knowledge, self-reverence, self-healing, and self-care processes while allowing increasing diversity.

This goal is pursued through the human-to-human caring process and caring transactions that respond to the subjective inner world of the person in such a way that the nurse helps individuals find meaning in their existence, disharmony, suffering, and turmoil and promotes self-control, choice, and self-determination with the health–illness decisions.

Nursing can contribute to the human sciences by establishing a set of values, assumptions, goals, and methods about humans and science that seek to integrate:

- the human mind, body, and soul as inseparable (as contrasted to a particulate view of the body)
- reality and fantasy
- facts and meaning
- objective and subjective worlds
- external and internal events
- disease, illness and health
- physical and metaphysical realms.

Society needs the caring professions, and nursing in particular, to help to restore humanity and nourish the human soul in an age of technology, scientism, loneliness, rapid change, and stresses, an age without moral or ethical wisdom, as to how to serve humanity.

This particular theory of nursing is metaphysical, in that it goes beyond the rapidly emerging existential–phenomenological approaches in nursing (for example, Paterson, Zderad, Parse, Taddy, and the important approaches of Rogers, Newman, King, and others) to a higher level of abstraction and a higher sense of personhood, which incorporates the concept of the soul and transcendence. The notion of a human soul is nothing new or original. It is, however, unusual to include it in a theory. The closest concept in psychology and nursing are concepts like self, inner self, "I," me, self-actualization, and so on. The bold attempt to acknowledge and try to incorporate a concept of the soul in a nursing theory is a reflection of an alternative position that nursing is now free to take. This new concept breaks from the traditional medical science model and is also a reflection of the scientific times. The evolution of the history and philosophy of science now allows some attention to metaphysical views that would have been unacceptable at an earlier point in time.

My basic beliefs and values about human life provide a foundation for my theory of nursing and become an integral part of nursing goals and nursing human care transactions, and influence the subject matter of the theory, the perspective on the subject matter, and the approach for reasoning.

Basic Premises

1. A person's mind and emotions are windows to the soul. Nursing care can be and is physical, procedural, objective, and factual, but at the highest level of nursing the nurses' human care responses, the human care transactions, and the nurses' presence in the relationship transcend the physical and material world, bound in time and space, and make contact with the person's emotional and subjective world as the route to the inner self and the higher sense of self.

2. A person's body is confined in time and space, but the mind and soul are not confined to the physical universe. One's higher sense of mind and soul transcends time and space and helps to account for notions like collective unconscious, causal past, mystical experiences, parapsychological phenomena, a higher sense of power, and may be an indicator of the spiritual evolution of human beings. (This idea has been proposed by numerous philosophers, including Teilhard de Chardin, Kierkegaard, Hegel, and Marcel.

3. A nurse may have access to a person's mind, emotions, and inner self indirectly through any sphere—mind, body or soul—provided the physical body is not perceived or treated as separate from the mind and emotions and higher sense of self (soul). This is consistent with Hippocrates who thought the person's mind and soul should be inspired before illness could be treated.

4. The spirit, inner self, or soul (geist) of a person exists in and for itself. The spiritual essence of the person is related to the human ability to be free, which is an evolving process in the development of humans. The ability to develop and experience one's essence freely is limited by the extent of others' ability to "be." The destiny of one's being (humankind's destiny) is to develop the spiritual essence of the self and in the highest sense, to become more Godlike. However, each person has to question his or her own essence and moral behavior toward others, because if people are dehumanized at a basic level, for example, a human care level, that dehumanizing process is not capable of reflecting humanity back upon itself.

5. People need each other in a caring, loving way. Love and caring are two universal givens. To paraphrase, Teilhard de Chardin "Love (and care) are the most universal, the most tremendous, and the most mysterious of cosmic forces. . . It is the primal and universal psychic energy."[3] These needs are often overlooked, or even though we know we need one another in a loving and caring way, we do

not behave well toward each other. If our humanness is to survive, we need to become more loving, caring, and moral to nourish our humanity, advance as a civilization, and live together. As a beginning we have to impose our own will to love, care, and be moral upon our own behavior, not on others' behavior. We need to love, respect, care for ourselves, and treat ourselves with dignity, before we can respect, love, and care for others and treat them with dignity.

6. A person may have an illness that is "completely hidden from our eyes." To find solutions it is necessary to find meanings. A person's human predicament may not be related to the external world as much as to the person's inner world as he or she experiences it.

7. The totality of experience at any given moment constitutes a phenomenal field. The phenomenal field is the individual's frame of reference and comprises the subjective internal relations and the meanings of objects, subjects, past, present, and future as perceived and experienced.

A PERSONAL ANECDOTAL NOTE FOR STUDENTS

Rose McKay suggested to me that the ideas represented by these values, goals, and beliefs lead to a prescriptive theory for nursing. That is, "if nursing assumes it now has the professional development, assumes it has a knowledge base and can assume competency, these ideas suggest we are now ready to review what we do with our knowledge, professional maturity and competences. As such they suggest a moral ideal and moral commitment."

These notions were posed to me in the spring of 1982 by Professor Rose McKay in her graduate theory class. The other issue raised by Dr. McKay was the overlap in my ideas between a philosophy and a theory. I am quite sure my ideas represent some of both, but not either exclusively.

In viewing my ideas as helping nursing and nurses to develop a meaningful philosophical base for one's practice and our science, and to examine what we stand for scientifically, socially, and morally, these do call for renewal and rededication in an age of professional confusion. I do not think of myself as having a prescriptive theory, and I even question whether nursing can truly have a prescriptive theory. Perhaps, however, the human values and moral ideals are prescriptive.

REFERENCES

1. Jung, C.G. Psychology and Alchemy. In H. Read, M. Fordham, & G. Adler (Eds.), The Collected Works of C.G. Jung (Vol. 12). Princeton, N. J.: Princeton University Press, 1968, 99–101.

2. Smith, J. The Idea of Health. New York: Teachers College, 1983, 31.
3. De Chardin, T. On Love. New York: Harper & Row, Pub., 1967, 7–8.

BIBLIOGRAPHY

De Chardin, T. On Love. New York: Harper & Row, Pub., 1967.
Jung, C.G. Psychology and Alchemy. In H. Read, M. Fordham, & G. Adler (Eds.), The Collected Works, (Vol. 12). Princeton, N. J.: Princeton University Press, 1968.
Smith, J. The Idea of Health. New York: Teachers College, 1983.

7

THEORY COMPONENTS AND DEFINITIONS

The human mind is forever moving. The activities of the mind have no limit, they form the surroundings of life. Surroundings have no more limits than the activities of the mind.

Baddyo Dendo Kyakai
The Teachings of Buddha*

This chapter is an extension of the previous chapter and includes more components of the theory, definitions of specific terms, and how they interrelate.

DEFINITION OF NURSING

Nursing, as a word, is a philosophical concept that suggests tenderness and holds various meanings for people. As such the concept of nursing is dynamic and changing. The word "nurse" is both a noun and a verb. There is the nurse as a person and the nurse as responses and behaviors. Nursing to me generally consists of knowledge, thought, values, philosophy, commitment, and action, with some degree of passion. The knowledge, values, action, and passion are generally related to human care transactions and intersubjective personal human contact with the lived world of the experiencing person.

---

*The Teachings of Buddha, 4th Ed. Tokyo, Japan: Kosaido Printing Co. Ltd., 1976

As such, human care and caring is viewed as the moral ideal of nursing. It consists of transpersonal human-to-human attempts to protect, enhance, and preserve humanity by helping a person find meaning in illness, suffering, pain, and existence; to help another gain self-knowledge, control, and self-healing wherein a sense of inner harmony is restored regardless of the external circumstances. The nurse is a coparticipant in a process in which the ideal of caring is intersubjectivity. Because of the human nature of nursing, the moral, spiritual, and metaphysical components of nursing cannot be ignored or replaced. They are inherently operating, directly or indirectly, and therefore, need to be acknowledged as part of a theorist's world view, belief system, and philosophy. In a sense, the metaphysical beliefs of a nursing theory provide the passion for nursing and keep it alive, changing, and open to new possibilities.

SCIENCE AND DISCIPLINE OF NURSING

Nursing in this context may be defined as a human science of persons and human health—illness experiences that are mediated by professional, personal, scientific, esthetic, and ethical human care transactions. Such a view requires the nurse to be a scientist, scholar, and clinician but also a humanitarian and moral agent, wherein the nurse as a *person* is engaged as an active coparticipant in the human care transactions. This science leans toward employing qualitative theories and research methods, such as existential–phenomenology, literary introspection, case studies, philosophical–historical work, and other approaches that allow a close and systematic observation of one's own experience and seek to disclose and elucidate the lived world of human health illness experience and the phenomena of human-to-human caring.

Since nursing science involves intersubjective human-to-human care, the process and practice of nursing become transpersonal and metaphysical. When these aspects of nursing are acknowledged and incorporated into our science, then nursing can cultivate a fuller access of the intuitive, esthetic, quasirational modes of thought, feeling, and action, and there can be greater use of our geist or spirit in relating to others from which the rational, Western scientific culture often closes us off.

This position does not discount scientific method or Western thought, rather it seeks to elucidate and acknowledge the other dimensions operating, if we are to understand the idea of the person, nursing, and human care processes in health and illness.

The *person* is viewed as "a being-in-the-world" and is the locus of human existence. The person exists as a living, growing gestalt. The person possesses three spheres of being—mind, body, and soul—that are influenced by the concept of self. The mind and the emotions are the starting point, the focal point, and the point of access to the body and soul. The notion of self is the subjective center that experiences and lives within the sum total of body parts, thoughts, sensations, desires, memories, life history, and so forth. The person is not simply an organism or material physical being; the person is also a part of nature, a spiritual being, neither purely physical, nor purely spiritual. A person's existence is embodied in experience, in nature, and in the physical world, but a person can also transcend the physical world and nature by controlling it, subduing it, changing it, or living in harmony with it.

A person is the experiencing and perceiving organism. The person and the self are the same when the person is congruent with the real self. That occurs when there is harmony within the mind, body, and soul of the person.

The *self* denotes the "organized consistent conceptual gestalt composed of perceptions of the characteristics of the 'I' or 'me' and the perceptions of the relationships of the 'I' or 'me' to others and to various aspects of life, together with the values attached to those perceptions. It is a fluid and changing gestalt, a process, but at any moment a specific entity."[1] One's self is a process; an unending process wherein new experience is turned into knowledge, each psychological moment shapes the next psychological moment. In addition to the self as it is; there is an ideal self that the person would like to be. The highest sense of the self connotes the spiritual self, the geist, soul, or the essence of the person's self with the "potential forms of consciousness entirely different from our waking consciousness."[2]

The totality of human experience (one's being-in-the-world) constitutes a *phenomenal field.** The phenomenal field is the individual's frame of reference that can be known only to the person. "It can never be known to another except through empathetic inference and then can never be perfectly known."[3] How a person perceives and responds in a given situation depends upon the phenomenal field (subjective reality) and not just upon the objective conditions or external reality.

A continuity of consciousness occurs over time. Each successive moment of awareness is shaped by the previous moment, and will determine the following moment. "The human is like a river that keeps a constant form, though not a single drop is the same as a moment ago."[4] A person's mental and emotional state and phenomenal field may vary from moment to moment as well as his or her sensory objects—sounds, smells, taste, sights, random memories, future plans; other thoughts mingle with and are often associated with objects of the senses and become part of the phenomenal field.

The phenomenal field is not identical with the consciousness but incorporates consciousness along with perceptions of self and others; feel-

---

* These ideas are influenced by writings on gestalt psychology and existential psychology by Carl Rogers, Kurt Goldstein, and Kurt Lewin, as well as by Eastern psychology.

ings, thoughts, bodily sensations, spiritual beliefs, desires, goals, expectations; environmental considerations; and meanings and the symbolic nature of one's perceptions—all of this is based upon one's life history and the presenting moment as well as the imaged future.

SPIRITUAL DIMENSION

The world of the spirit and soul becomes increasingly more important as a person grows and matures as an individual and as humankind evolves collectively. The salience of the spiritual aspect of a person or race varies from individual to individual, from culture to culture, and within them. As William James said, "Our normal waking consciousness is but one special type of consciousness, whilst all about it, parted from it by the filmiest of screens, lies potential forms of consciousness entirely different."[5] Some cultures can be considered more spiritually evolved than others. For example, the Eastern cultures of India and Egypt, with their long histories of spiritual valuing, can be viewed as more spiritually developed and having a greater capacity for higher levels of consciousness than our Anglo-Saxon, Western world, with its values on physical-materialism combined with the Western world's relatively short history. There is evidence, however, that the Western world's values are moving toward Eastern spiritualism. This movement is manifest in the increase of Eastern philosophies and ideas that are incorporated into health programs, yoga, exercise, health foods, fasting, meditation, and special diets. The works of Gardner Murphy and Lois Murphy in Asian psychology report that the psychological interests of the East and West are quickly coming together.[6]

The *world* refers to all those forces in the universe, as well as a person's immediate environment and situation that affect the person, be they internal, external, human, humanmade, artificial, natural, cosmic, psychic, past, present or future.

Harmony-Disharmony
Where there is disharmony among the mind, body, and soul or between a person and the world, there is a disjunctive between the self as perceived and one's actual experience, and there is also a felt incongruence within the person, between the I and me and between the person and the world. Incongruence between the self as perceived and a person's experience reflects the presence of disharmony within the mind, body, and soul—the I is not equal to the real self or the real me. This incongruence leads to threat, anxiety, inner turmoil, and can lead to a sense of existential despair, dread, and illness. If prolonged, it can contribute to disease.

If there is harmony within a person's mind, body, and soul then a sense of congruence will exist between the I and me; between the self as perceived and the self as experienced by the person.

Another element of congruence–incongruence is the congruence—or lack of it—between subjective reality (the phenomenal field) and external reality (the world as it is). An additional degree of congruence–can occur between the self and the ideal self. If a person does not feel congruent with the mind, body, and soul, for example, rejects the self as it is, or is obsessed with an ideal self, the person will be dissatisfied and maladjusted. Moreover, incongruence can occur if there is a lack of union with another human or if a person feels separate and alone in his or her quest to be and grow. Another type of incongruence can occur if there is a lack of harmony between a person and nature; such a position calls attention to the need for nature and esthetics in one's world.

Striving

The person has one basic striving: to actualize the real self, thereby developing the spiritual essence of the self, and in the highest sense, to become more Godlike. In addition, each person seeks a sense of harmony within the mind, body, and soul and thereby further integrates, enhances, and actualizes the real self. The more one is able to experience one's real self, the more harmony there will be within the mind, body, and soul and a higher degree of health will exist. Because disharmony is associated with illness, and harmony is associated with health, the nursing profession needs to be concerned with how disharmony develops and how the self and real self can be made more congruent, thereby establishing harmony with the mind, body, and soul. The nurse caring transaction contributes to the person's developing spiritual essence, or real self, and likewise contributes to more self-knowledge, self-reverence, self-control, and self-healing for both the nurse and patient.

Human behavior is basically the goal-directed attempt of the person to satisfy needs as experienced in the perceived phenomenal field. Although there are many needs, each of them is subservient to basic striving toward actualizing one's spiritual self and establishing harmony within the mind, body, and soul.

In early life the person is more attentive to a sense of harmony between mind/emotions and body, but as one grows and allows for the existential and spiritual side of one's self to develop, the person becomes more concerned with incongruence or disharmony between the self and the real self.

As a person matures he or she becomes more differentiated and his or her sense of inner self becomes more developed. The person then seeks a greater degree of harmony with his or her soul because of a higher sense of discrimination.

Human needs consist of the need to be loved and cared for and about, the need for positive regard and the need to be accepted, understood, and valued.[7] There is also a human need to achieve union, transcend one's individual life, and find harmony with life.

Transpersonal human care and caring transactions are those scientific, professional, ethical, yet esthetic, creative and personalized giving–receiving behaviors and responses between two people (nurse and other) that allow for contact between the subjective world of the experiencing persons (through physical, mental, or spiritual routes or some combination thereof).* The human care transactions include the nurse's unique use of self through movements, senses, touching, sounds, words, colors, and forms in which he or she transmits and reflects the person's condition back to that person. He or she does this in such a way that allows for the release and flow of his or her intersubjective feelings and thoughts, and pent up energy. Such a transaction, in turn, helps to restore inner harmony while also contributing to the patient and nurse finding meaning in the experience. However, the origin of the meaning for both resides within rather than existing without. In this process the nurse is also attending to a concern above all for the dignity of the person as an important end. Dignity in this sense "simply expressed as being has dignity when it gives to itself, its meaning and so creates for itself integrity."[8]

The contact with the subjective world has the potential to go beyond bodily or mental–emotional contact or interaction, and to reach out and touch the higher, spiritual sense of self, or the soul. As such, transpersonal human care occurs from person to person in an *I–Thou* relationship. It can release inner power and strength and help the person gain a sense of inner harmony within the mind, body, and soul; this contact and process in turn generates and potentiates the self-healing processes.

The two individuals (the nurse and the other) in a caring transaction are both in a process of being and becoming. Both individuals bring with them to the relationship a unique life history and phenomenal field, and both are influenced and affected by the nature of the transaction, which in turn becomes part of the life history of each person. In this sense of a caring transaction, caring is a moral ideal, rather than an interpersonal technique and it entails a commitment to a particular end. The end is the protection, enhancement, and preservation of the person's humanity, which helps to restore inner harmony and potential healing.

An Event or Actual Caring Occasion

Two persons (nurse and other) together with their unique life histories and phenomenal field in a human care transaction comprise an *event*.†

---

* Transpersonal refers to an intersubjective human-to-human relationship in which the person of the nurse affects and is affected by the person of the other. Both are fully present in the moment and feel a union with the other. They share a phenomenal field which becomes part of the life history of both and are coparticipants in becoming in the now and the future. Such an ideal of caring entails an ideal of intersubjectivity, in which both persons are involved.

†Based on Whitehead's notion of EVENT; actual occasion.[9]

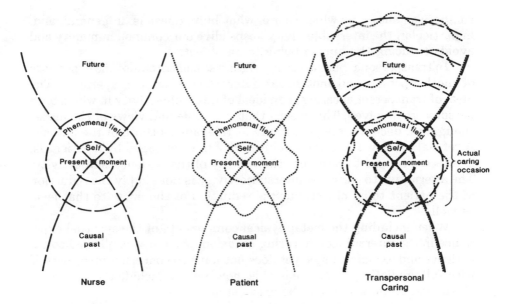

Figure 5. Dynamics of human caring process, including nurse-patient transpersonal dimension. (*Illustration by Mel Gabel, University of Colorado, Biomedical Communications Dept.*)

An event, such as an actual occasion of human care, is a focal point in space and time from which experience and perception are taking place, but the actual occasion of caring has a field of its own that is greater then the occasion itself. As such, the process can go beyond itself, yet arise from aspects of itself that become part of the life history of each person, as well as part of some larger, deeper, complex pattern of life. (Fig. 5)

An *actual caring occasion* involves action and choice both by the nurse and the individual. The moment of coming together in a caring occasion presents the two persons with the opportunity to decide how to be in the relationship—what to do with the moment. Whatever is decided involves one manner and not another. If the caring occasion is indeed transpersonal and allows for the presence of the geist or spirit of both, then the event expands the limits of openness and has the ability to expand the human capacities. It thereby increases the range of certain events that could occur in space and time at the moment as well as in the future. The moment of the caring occasion becomes part of the past life history of both persons and presents both with new opportunities. Such an ideal of intersubjectivity between the nurse and patient is based upon a belief that we learn from one another how to be human by identifying ourselves with others or finding their dilemmas in ourselves. What we all learn from it is self-knowledge. The self we learn about or discover is every self: it is universal—the human self. We learn to recognize ourselves in others. The comparison shows us what we are, what humanness is, in general, and, in particular, the intersubjectivity keeps alive our common humanity and avoids reducing the human being to an object.

In transpersonal human caring, the nurse can enter into the experience of another person, and another can enter into the nurse's experience. The ideal of transpersonal caring is an ideal of intersubjectivity in which both persons are involved. This means that the value and views of the nurse, though not decisive, are potentially as relevant as those of the patient. A refusal to allow the nurse's subjectivity to be engaged by a patient is, in effect, a refusal to recognize the validity of the patient's subjectivity. The alternative to caring as intersubjectivity is not simply the reduction of the patient to an object, but the reduction of the nurse to that level as well."[10]

When including the metaphysical components of the spiritual experience in the intersubjective caring occasion, the nurse is allowed to experience and explain things, yet does not have to concern him- or herself with full prediction; he or she is able, however, to include the mysteries of life and unknowns yet to be discovered.

An actual caring occasion is located not only in the simple physical instance of a given moment of time, but the event/experience has internal relations to other objects–subjects in the phenomenal field plus internal subjective relations between the past, present, and imagined future for each person and for the whole. An actual caring occasion can be present in the life of both the nurse and person beyond the physical instance of the given point in time.

Time
One cannot clearly distinguish between past and present time even though the present is more subjectively real and the past is more objectively real. The past is prior to, or in a different mode of being than the present, but it is not clearly distinguishable. Past, present, and future instants merge and fuse.

The notion of the moment, actual occasion, and time is depicted by Virginia Woolf in *A Writer's Diary.*[11]

> Friday, January 4th. "Now is life very solid or very shifting? I am haunted by the two contradictions. This has gone on forever; will last forever; goes down to the bottom of the world—this moment I stand on. Also it is transitory, flying, diaphanous. I shall pass like a cloud on the waves. Perhaps it may be that though we change, one flying after another, so quick, so quick, yet we are somehow successive and continuous we human beings; and show the light through."

Figure 5 depicts the various components of transpersonal caring, including self, phenomenal field, and actual caring occasion of the patient and nurse, intersubjectively coming together in a given moment in time whereby the past, present, and future merges.

The actual caring occasion in a presenting moment has the potential to influence both the nurse and patient in the future. The caring occasion then becomes part of the subjective, lived reality and the life history of both. Both are coparticipants in becoming in the now and the future, and both are part of some larger, deeper, complex pattern of life.

REFERENCES

1. Rogers, C.R. A theory of therapy, personality, and interpersonal relationships, as developed in the client-centered framework. In S. Koch, (Ed.), Psychology: A Study of a Science, (Vol. 3). New York: McGraw-Hill, 1959, 200.
2. James, W. The Principles of Psychology. New York: Dover, 1950, 305.
3. Rogers, C.R. A theory of therapy, personality, and interpersonal relationships, as developed in the client-centered framework. In S. Koch, (Ed.), Psychology: A Study of a Science, 210.
4. Van Aung, Z. (Ed. and trans.) Compendium of Philosophy. London: Pali Text Society, 1972, 7.
5. James, W. The Principles of Psychology, 305.
6. Murphy, G. & Murphy, L. B. (Eds.). Asian Psychology. New York: Basic Books, 1968.
7. Watson, J. Nursing: The Philosophy and Science of Caring. Boston: Little, Brown, 1979, 183–193.
8. Gadow, S. Existential advocacy as a form of caring: Technology, truth, and touch. Paper presented to the Research Seminar Series: The Development of Nursing as a Human Science. The School of Nursing, University of Colorado Health Sciences Center, Denver, March 1984.
9. Whitehead, A.N. Science and the Modern World. Cambridge: Cambridge University Press, 1953.
10. Gadow, S. Existential advocacy as a form of caring: Technology, truth, and touch.
11. Woolf, V. A writer's diary. In J. Hersey, (Ed.), The Writer's Craft. New York: Knopf, 1975, 374.

BIBLIOGRAPHY

Gadow, S. Existential advocacy as a form of caring: Technology, truth, and touch. Paper presented to the Research Seminar Series: The Development of Nursing as a Human Science. The School of Nursing, University of Colorado Health Sciences Center, Denver, March 1984.

Guenther, H.V. Philosophy and Psychology in the Abhidhamma. Berkeley, Calif.: Shambhala, 1976.

James, W. The Principles of Psychology. New York: Dover, 1950.

Murphy, G. & Murphy, L.B. (Eds.). Asian Psychology. New York: Basic Books, 1968.

Rogers, C.R. A theory of therapy, personality, and interpersonal relationships, as developed in the client-centered framework. In S. Koch (Ed.) Psychology: A Study of a Science, (Vol. 3). New York: McGraw-Hill, 1959.

Van Aung, Z. (Ed. and trans.). Compendium of Philosophy. London: Pali Text Society, 1972.

Watson, J. Nursing: The Philosophy and Science of Caring. Boston: Little, Brown, 1979.

Whitehead, A.N. Science and the Modern World. Cambridge: Cambridge University Press, 1953.

Woolf, V. A writer's diary. In J. Hersey, (Ed.), The Writer's Craft. New York: Knopf, 1975.

8

TRANSPERSONAL CARING RELATIONSHIP

A transpersonal caring relationship connotes a special kind of human care relationship—a union with another person—high regard for the whole person and their being-in-the-world. Caring, in this sense, is viewed as the moral ideal of nursing where there is the utmost concern for human dignity and preservation of humanity. Human care can begin when the nurse enters into the life space or phenomenal field of another person, is able to detect the other person's condition of being (spirit, soul), feels this condition within him- or herself, and responds to the condition in such a way that the recipient has a release of subjective feelings and thoughts he or she had been longing to release. As such, there is an intersubjective flow between the nurse and patient.

As feelings, thoughts, and energies that are less harmonious with either person's self are released, they become replaced by other feelings, thoughts, and energies that are more harmonious with one's self and are kinder toward and more mindful of the well-being of each person and ultimately for humankind.

The simple, yet complex, human-to-human care process in nursing is a very basic foundation and starting point in which a harmonizing transpersonal caring relationship can occur.

A transpersonal caring relationship depends upon:

1. A moral commitment to protect and enhance human dignity, wherein a person is allowed to determine his or her own meaning.

2. The nurse's intent and will to affirm the subjective significance of the person (I–Thou versus I–It).
3. The nurse's ability to realize and accurately detect feelings and the inner condition of another. This can occur through actions, words, behaviors, cognition, body language, feelings, thought, senses, intuition, and so on.
4. The ability of the nurse to assess and realize another's condition of being-in-the-world and to feel a union with another. The nurse must be able to express the condition and/or attend to the condition through various means, such as movement, gestures, looks, acts, procedures, information, touch, sound, words, color, and form and other similar scientific, esthetic, and human means. The subjectivity of the patient is assumed to be as whole and as valid as that of the nurse. Mutuality, therefore, is a moral foundation of nursing. (If conditions 1 through 4 are present, the recipient is better able to have a release of some of the disharmony, of the mind, body, and soul and be freer to direct pent up energy to his or her own healing process.)
5. The nurse's own life history (casual past) and previous experiences and opportunities of having lived through or experienced one's own feelings and various human conditions, and of having imagined others' feelings in various human conditions. (Such knowledge and sensitivity can be gained through working with other cultures, the study of the humanities, art, drama, literature, exploring values, and having a good relationship with self. It can also be facilitated through values clarification, personal psychotherapy, meditation, autogenic control mechanisms, and so on. It is related to personal growth, maturity, and development of the nurse's self, a sensitivity to the self and others, and a human value system.)

USE OF ENTIRE SELF IN NURSING

The introduction of the professional nurse as a person in a transpersonal relationship with the patient may conflict with traditional views of the professional nurse. Nurses and other health professionals have been warned to avoid personal interactions, and personal involvement is considered unprofessional. Sally Gadow makes a convincing case for softening the distinction between the person and the professional.[1] Her ideas are consistent with caring transactions developed here and the notion of transpersonal caring.

Gadow emphasizes that even if responding person to person there will still be important differences between the nurse and patient, although it is still possible to allow the "amount" of the personal involvement to be equal. The concept of professional involvement as the participation of the entire self, using every dimension of the person as a resource in the professional relation, is entailed in the concept of a transpersonal caring relationship between the nurse and person.

Some of the personal differences have been classified in terms of (1) focus, (2) intensity, and (3) perspective.[2]

The *focus* of a patient's personal involvement in a professional relationship, for example, is directed toward the problem at hand and its effect upon his or her life. The concern is unavoidably self-directed. In contrast, the personal involvement of the nurse/person is directed away from one's own self toward the other's self. The nurse's feelings may be experienced and felt, not as a way of obtaining relief or help from patients, but as part of being and becoming in the relationship.

Personal relationships between friends have a give and take process of exchange wherein the one who needs the most receives from the other. However, there is a reciprocal sharing and an accepted norm of mutuality where each party in the relationship helps the other. A professional relationship may and does allow for the nurse person to benefit and be influenced by the other; however, the nurse does not depend upon receiving from the patient to maintain the involvement.

The *intensity* of the relationship that Gadow outlines is also experienced differently by the nurse and the patient. The patient experiences the immediacy of the distress. Intensity and likewise, immediacy, may indeed be felt by the nurse, but they serve to generate the reflective process necessary for help to be given. "Being able to help has greater value than simply sharing the other's experience. . . . This is in order to integrate feelings and knowledge in the attempt to alleviate the patient's distress."[3]

Lastly, the *perspectives* of the two persons (nurse and patient) differ. The nurse is externally involved, whereas the patient feels the pain from the inside and knows that it is his or her body, mind, and, in this instance, soul that is affected.

As Gadow goes on to explain, the difference in perspectives between the two is often used to indicate degrees of emotional involvement. The implication is that the patient is more emotional and the nurse less emotional. There is the real possibility, however, that both persons might (and do) experience emotional intensity.

The professional nurse differs from the patient or a friend in that the nurse helps integrate the subjective experience and emotions with the objective, external view of the situation. Even acknowledging some differences between personal and professional relations most assumed conditions related to unequal involvement do not exist. For example, the nurse does not manifest less involvement as a person than does the patient. The form and direction of the involvement may differ, but the amount of the nurse's involvement is equally as great. Personal involvement in a professional capacity is not an alternative to other kinds of involvement. It is a synthesis of involvements, a participation of the entire self, using every dimension of the person as a resource in the professional relation.

Because nursing is so immediate, and often intimate and private, with scientific, artistic, humanistic, ethical, and technical complexities, it offers avenues for every dimension—of the professional including the emotional, mental, esthetic, intuitive, physical, spiritual, and experiential—to be involved.

At many points throughout life, people are confronted with existential concerns about their existence and the meaning of their life. These concerns tend to be more urgent when the person's existence is threatened. This can occur from a threat to body, mind, inner self, and soul. Indeed a threat to any one of these aspects of a person in turn affects the others. Decisions are made about one's existence by questioning how one lives one's life, one's priorities, forms of coping, self-caring behaviors, different health and medical practices, degree of help or freedom desired, and so forth.

When the question of meaning arises, people can generally benefit by determining the meaning of their experience. Ideally, a person should have the opportunity for self-determination of the meaning of a health–illness experience before professionals make decisions about treatment or intervention.

The patient has opinions and meanings attached to the health–illness experience and can be free to examine that meaning and have that meaning incorporated into the professional's response to the situation. Nursing's continuity with persons often enables the nurse "to experience individuals as unique human beings continuously engaged in creating their own histories."[4] It is in this sense of the nurse–patient relationship that nursing meets the conditions for caring referenced earlier.

The conflict in nursing and in studying and developing ideas of human caring is related to the question of materialism versus spiritualism. The theory of nursing and transpersonal caring is based upon the premise that nursing needs "to develop a scientific tradition related to the vast uncharted (by science) sea of human potentials we call spiritual potentialities."[5]

The perspective, values, and subject matter developed herein are one attempt to acknowledge and actualize the personal and individual spiritual potentials that can expand through a transpersonal caring relationship. When the natural self of the nurse and patient coparticipate in a caring transaction, it potentiates self-healing human integrity and there is greater harmony for both nurse and person.

In a transpersonal caring relationship, a spiritual union occurs between the two persons, where both are capable of transcending self, time, space, and the life history of each other. In other words, the nurse enters into the experience (phenomenal field) of another and the other person enters into the nurse's experience. This shared experience creates its own phenomenal field and becomes part of a larger, deeper, complex pattern of life.

THE ART OF TRANSPERSONAL CARING

According to Tolstoy, if one ponders the meaning of art, one must first consider that art is a condition of human life and a means of human-to-human contact.

The human-to-human contact causes the receiver of art to enter into a certain kind of relationship both with the producer of art and with all those who simultaneously, previously, or subsequently receive the same or similar artistic expression. Another special feature of art is that feelings are transmitted.

The activity of art is based on the fact that a person receiving another person's expression of feeling, through hearing, sight, or even intuition, is capable of experiencing the emotion that moved the other to express it. To use the simplest example: one person laughs, and another who hears becomes merry, or a person weeps, and another who hears feels sorrow. If a man is excited or irritated, another who sees him may develop a similar state of mind. By his movements or by the sounds of his voice, a person expresses courage and determination, anger or calmness, and this state of mind passes on to others. A person suffers, manifesting his suffering by groans and spasms, and this suffering transmits itself to other people; a man expresses his feelings of admiration, devotion, fear, respect, or love of certain persons, objects, or phenomena, and others are moved by the same feelings of admiration, devotion, fear, respect, or love, to the same persons, objects, or phenomena.

It is on this capacity of one human being to receive another human being's expression of feeling and to experience those feelings for oneself that the artistic activity of nursing and caring is based.

The art of caring in nursing begins when the nurse, with the object of joining another (or others) to oneself with a certain feeling of care and concern, expresses that feeling by certain external indications.

Art is not just moving another directly and immediately by giving vent to feelings at the very time they are experienced. If a person causes another to relax, cry, or laugh when he himself is obliged to laugh, or cry, that does not amount to the art of caring. It is art when the nurse, having experienced or realized the feelings of another, is able to detect and sense these feelings and in turn is able to express them in such a way that the other person is able to experience them more fully and release the feelings he or she has been longing to express.

The feelings that the nurse as an artist inculcates in others may be varied—strong or weak, important or insignificant; they may be feelings of love, fear, courage, joy, quietness, awe. It is all art.

The activity of the art of caring in nursing is triggered by the human interaction in a nursing care situation. The human interaction evokes within the nurse a feeling. Having experienced that feeling, and having evoked it in oneself, then by means of movements, touch, sounds, words, colors, and forms, the nurse transmits the feeling so that another person experiences the same feeling—that is the activity of the art of caring.

The art of caring in nursing is a human activity consisting of the following: a nurse consciously, by means of certain signs, passes on to others feelings he or she has lived through, realized, or learned; others are united by these feelings and also experience them. The chief peculiarity of this feeling is that the recipient of a truly artistic caring impression is so united to the nurse's expression that he or she feels the feeling as his or her own, to the degree that what is expressed is just what he or she had been longing and wishing to express. A truly caring nurse/artist is able to destroy in the consciousness of the recipient the separation between him- or herself and the nurse. In other words, the nurse is able to form a union with the other person on a level that transcends the physical, and that preserves the subjectivity and physicality of persons without reducing them to the moral status of objects. As such, there is a freeing of both persons from their separation and isolation, in this uniting, of the feeling with another and with others who have also experienced the same feelings. The union of feelings can potentiate self-healing and discovery of inner power and control, and contribute to another finding meaning in his or her own existence. That is the great attractive force of the art of transpersonal caring in nursing.

If the nurse is able to detect accurately the other's condition of soul, if he or she feels this emotion, and this union with another, and in turn can express it accurately, then the recipient has a release of the feeling he or she has been longing and wishing to express; then human subjectivity is restored. This is what I call the art of transpersonal caring. If, however, there has been no such awareness, no union of the feelings of the moment which are moved by the human caring process, then transpersonal caring has not occurred.

I view this act of transpersonal caring in nursing as a human art, a human science, and a moral ideal of nursing.

Premises

The more individual the feelings are that the nurse transmits, the more strongly does the caring process affect the recipient. The more individual the state of soul is that is transferred, the more release, satisfaction, pleasure, and peace does the recipient obtain, and the more readily and strongly does another experience the caring and join in it. In other words, the full use of the self is called upon in this level of nursing.

A clarity of expression assists the nurse's caring. The more clearly the feeling is transmitted (which the other has long known and felt at some level, but which is now realized more fully, and for which expression has only now been found), the better able is the recipient to experience the union.

Most of all, the degree of transpersonal caring (in this sense of unity of feeling) is increased by the degree of genuineness and sincerity of the nurse. If the recipient feels that the nurse is contriving the feelings and has not actually felt a union with the condition of the other's soul but is "going through the process," trying to act upon the other's feelings, and does not feel within his or her self what longs to be expressed, resistance immediately springs up. Thereafter, the most unique response, elaborate approach, and the cleverest techniques not only fail to release, but actually repel another person and contribute to the state of disharmony (illness).

The sincerity and the individuality go together, because if the nurse is sincere he or she will be able to express the feeling as he or she experiences or realizes it. Since every nurse is a unique person, his or her feelings will be individual. The more individual they are, the more the nurse as artist has drawn them from the depths of his or her nature, and the more natural and genuine they will be. The condition of genuineness and sincerity has always been complied with in peasant art and nature, and this helps explain why such acts are so powerful.

However, it is a condition that is almost entirely absent in our educational and socializing experiences in the nursing profession and is instead substituted with artificial aims of professionalism and scientism. Such are the conditions of transpersonal caring that help to decide the artistic expression of the human care process in nursing, which as a moral ideal is considered apart from the facts of the subject.

Each one of the above mentioned conditions may vary. For one nurse, for example, the level of spirituality and individuality of the feeling transmitted may predominate; for another nurse, clarity of expression; for a third, sincerity and genuineness; while a fourth may have sincerity and individuality but be deficient in clarity; a fifth may have individuality and clarity, but less sincerity; and so forth, to all possible degrees and in all combinations. The union between the nurse and recipient of care allows for both to have an intimate relationship with their spiritual selves.

Likewise, the nurse, through spirituality, individuality, clarity, and sincerity may draw upon various combinations of expression of feelings. One may rely more on acts and movements; another on words and sounds; another on silence and nonverbal expression; and still another on precision, forms or color, and so forth, again to all possible degrees and in all possible combinations. Thus is it true for the nurse, and thus is it true for the recipient regarding the possible degrees and combinations with which the one communicates the condition of the soul.

SUMMARY OF TRANSPERSONAL CARING

The art of transpersonal caring in nursing as a moral ideal is a means of communication and release of human feelings through the coparticipation of one's entire self in nursing. Transpersonal caring, therefore, is a means of progress where an individual moves toward a higher sense of self and harmony with his or her mind, body, and soul.

Collectively, the art of transpersonal caring allows humanity to move towards greater harmony, spiritual evolution, and perfection. Such a union of feelings and caring communication renders accessible to humans all the knowledge discovered by the experience of and reflection upon preceding generations as well as people in one's own time. The art of transpersonal caring in nursing makes accessible to a person a sense of humanity and intersubjectivity experienced by previous individuals and contemporaries in similar human conditions.

As the evolution of human thought progresses with the addition of truer and more necessary knowledge, dislodging and replacing what was mistaken and unnecessary, so the evolution of human feelings can proceed by means of the moral ideal of transpersonal caring in nursing, whereby feelings less kind and less necessary for the well-being of humankind can be replaced by kinder and more necessary feelings for the well-being and dignity of humankind. This human process is the value of the art of transpersonal caring in nursing.

The more the art of transpersonal caring in nursing advances the kinder and more helpful feelings for the human, the more we can define ideal caring with reference to its content and subject matter of nursing.

In the idea of transpersonal nursing, as developed through the study of Tolstoy's work, there is room for art, science, ethics, and metaphysics. With its attention to the human process, care activity, the intersubjective feelings, the individuality of each nurse and recipient of care, the process allows for combinations of expressions of human feelings in different moments and contexts, and with different outcomes, that can never be fully explained or predicted.

The emphasis on the human component of the feelings and the human-to-human caring is consistent with my view of the *person*. Transpersonal caring not only allows for release of emotions and the evolution of the person's spiritual self or soul, but it promotes congruence between the person's perception and experience, and promotes self as is and ideal self, and harmony within the person's mind, body, and soul. The process allows the nurse to reflect the self back upon the self.

The transpersonal caring process is largely art because of the way it touches another person's soul and feels the emotion and union with another, the goal being the movement of the person toward a higher sense of self and a greater sense of harmony within the mind, body, and soul. In turn, the process contributes to the movement of humans toward perfection, and contributes to the preservation of humanity.

The union of the two persons in which the condition of the human soul and feelings have been transmitted allows for liberation of the human mind and soul that leads to a greater sense of strength, power, and human capacity for finding the meanings in existence and illness. It also establishes an intimate relationship with one's spiritual self, through the artistic human caring transactions.

The transpersonal caring of the nurse in which the feelings are released allows the recipient of care to assimilate better the condition of one's soul into his or her self. The assimilation may lead to some reorganization of one's self as perceived and one's self as experienced. Both will *be* in a more unified way what each spiritually *is* in essence.

The context for viewing the person and the nurse in the theory of transpersonal caring is in the moment-to-moment human encounters between the two people. The coming together of the two—one the care-giver and the other the recipient—comprise an event. An event connotes an actual caring occasion in which intersubjective caring transactions occur.

The person's mind and emotions then become the starting point, the focal point, and the point of access to the soul and the body. Once the person moves toward a higher sense of self with increased harmony, then one's own self-healing processes and one's capacity for finding meaning in existence are available. At that point one can better choose between health and illness, regardless of any disease, bodily or human condition.

It is not so much the *what* of the nursing acts, or even the caring transaction per se, it is the *how* (the relation between the *what* and the *how*), the transpersonal nature and presence of the union of two persons' soul(s), that allow for some unknowns to emerge from the caring itself.

The ideas expressed here about transpersonal caring as a moral ideal for nursing allow nurses to call upon the inner depth of their own humanness and personal creativity as they realize the conditions of a person's soul and their own. The intersubjective process of caring is infinite and will continue to expand as knowledge and approaches expand. The transpersonal human-to-human caring is the essence and moral ideal of a style of nursing where human dignity and humanity are preserved and human indignity is alleviated in health–illness experiences.

REFERENCES

1. Gadow, S. Existential advocacy: Philosophical foundation of nursing. In S. Spicker & S. Gadow, (Eds.), Nursing Images and Ideals. New York: Springer, 1980, 86-101.
2. Ibid., 87-92.
3. Ibid., 88, 89.
4. Ibid., 98.
5. Tart, C. (Ed.). Transpersonal Psychologies. New York: Harper & Row, Pub., 1976, 58.
6. Tolstoy, L. [What is art?] (L. & A. Maude, trans.) In J. Hersey, (Ed.), The Writer's Craft. New York: Knopf, 1975, 25-30.

BIBLIOGRAPHY

Gadow, S. Existential advocacy: Philosophical foundation of nursing. In S. Spicker & S. Gadow (Eds.), Nursing Images and Ideals. New York: Springer, 1980.
Tart, C. (Ed.). Transpersonal Psychologies. New York: Harper & Row, Pub., 1976.
Tolstoy, L. [What is art?] (L. & A. Maude, trans.) In J. Hersey, (Ed.), The Writer's Craft. New York: Knopf. 1975.

9

STRUCTURAL OVERVIEW OF WATSON'S THEORY OF HUMAN CARE

SYNOPSIS

Subject Matter
The primary subject matter of this theory includes:

1. Nursing within a human science and art context.
2. Mutuality of person/self of both nurse and patient with mind-body–soul gestalt, within a context of intersubjectivity.
3. Human care relationship in nursing as a moral ideal that includes concepts such as phenomenal field, actual caring occasion, and transpersonal caring.

Also inherent within the subject matter are the notions of health–illness, environment, and universe, and how they interact, transact, and can transcend the physical–material objects and values of life.

Values
The values inherent in this work are associated with deep respect for the wonder and mysteries of life and the power of humans to change; a high regard and reverence for the spiritual–subjective center of the person with power to grow and change; a nonpaternalistic approach to helping a person gain more self-knowledge, self-control, and self-healing, regardless of the presenting health–illness condition. This value system is blended with carative factors[1] such as humanistic–altruism, sensitivity to self and others, and a love for a trust of life and other humans. Caring is presented

as a moral ideal of nursing with a concern for preservation of humanity, dignity and fullness of self.[2]

Goals

The goals for the theory ideals are associated with mental–spiritual growth for self and others, finding meaning in one's own existence and experiences, discovering inner power and control, and potentiating instances of transcendence and self-healing.

Agent of Change

The agent of change in this work is viewed as the individual patient, but the nurse can be a coparticipant in change through the human care process. The agent of change is not the physician, nurse, medication, treatment, or technology per se, but the personal, internal mental–spiritual mechanisms of the person who allows the self to be healed through various internal or external means, or without external agents, but through an intersubjective interdependent process wherein both persons may transcend self and usual experiences. Such a view holds a commitment to a particular end, beyond disease or pathology per se, but a moral ideal toward preservation of harmony with mind–body–soul and maintaining human dignity and integrity.

Interventions

The interventions in this theory are related to the human care process with full participation of the nurse/person with the patient/person.* Human care requires knowledge of human behavior and human responses to actual or potential health problems;[3] knowledge and understanding of individual needs; knowledge of how to respond to others' needs; knowledge of our strengths and limitation; knowledge of who the other person is, his or her strengths and limitations, the meaning of the situation for him or her; and knowledge of how to comfort, offer compassion and empathy. Human care also requires enabling actions, that is, actions that allow another to solve problems, grow, and transcend the here and now, actions that are related to general and specific knowledge of caring and human responses.

The interventions related to the human care process require an intention, a will, a relationship, and actions. This process entails a commitment to caring as a moral ideal directed toward the preservation of humanity. The process affirms the subjectivity of persons and leads to positive

---

* *Interventions* is used in categorizing nursing theory and components of a nursing model in the literature. The term intervention sounds harsh and mechanical and is inconsistent with my ideas and ideals. A more consistent term for my purposes might be caring process. For pedagogical purposes, however, I have retained the term interventions here.

change for the welfare of others, but also allows the nurse to benefit and grow. The combination of interventions can be referred to as carative factors* and include[4]:

1. Humanistic–altruistic system of values
2. Faith–hope
3. Sensitivity to self and others
4. Helping–trusting, human care relationship
5. Expressing positive and negative feelings
6. Creative problem-solving caring process
7. Transpersonal teaching–learning
8. Supportive, protective, and/or corrective mental, physical, societal, and spiritual environment
9. Human needs assistance
10. Existential–phenomenological–spiritual forces.

All of this is presupposed by a knowledge base and clinical competence.

All of these carative factors become actualized in the moment-to-moment human care process in which the nurse is being with the other person (whether administering an emergency intravenous treatment to a critical care patient or changing the linen of an unconscious patient). Human care requires the nurse to possess specific intentions, a will, values, and a commitment to an ideal of intersubjective human-to-human care transaction that is directed toward the preservation of personhood and humanity of both nurse and patient. Although these are ideals, different nurses and different moments allow higher levels of caring. The degree of caring is influenced by multiple, complex forces. The more human care is actualized as an intersubjective moral ideal, in each moment-to-moment caring occasion, the more potential the caring holds for human health goals to be met through finding meaning in one's own existence, discovering one's own inner power and control, and potentiating instances of transcendence and self-healing. Lastly, the human science, human care theory allows nursing individually and collectively to contribute to the preservation of humanity in an individual in society: the moral ideals and intersubjective human processes proposed here for nursing foster the spiritual evolution of humankind.

Perspective
The perspective is spiritual–existential and phenomenological in orientation, but also draws upon some Eastern philosophy.

---

*From Watson's work *Nursing: The Philosophy and Science of Caring.* Boston: Little, Brown, 1979. Slight changes are made in the language of the carative factors, especially numbers 6 and 10.

Context

The context is humanitarian and metaphysical. It incorporates both the art and science of nursing. Science is emphasized in a human science context.

Approach

The approach is descriptive, but the moral ideals have been described as possibly prescriptive (see p. 51).

Method

The optimal method for studying the theory is more naturally through field study that is qualitative in design. My views are most congruent with a phenomenological–existential methodology for study and inquiry. The ideas also allow for an applied humanities or applied philosophy or ethics approach. A word of caution regarding method is that there is no consensus regarding *one* scientific method. It depends on what components of the theory one chooses to research. The best rule of thumb is that the method should fit the phenomena under study; one should not force a phenomenon of interest into a scientific method when there is an acknowledged need for alternative or new methods, and vice versa. I am supportive of a range of methods, but would encourage scholars to pursue a creative-paradigm transcending approach that fits with a new philosophy of science, but maintain high standards, rigor, and credibility. Examples are historical research, comparative case studies, photographic–artistic documentation, literary works, philosophical analysis, and subjecting clinical data to new analytic techniques, such as empirical phenomenological analysis, neo-ethnographic approaches, and so on.

The method is applied to some of my research data. The same data are presented from the two different methodological perspectives related to phenomenology. Such an approach can perhaps help the reader to appreciate the different approaches and recognize why such methods are conceptually consistent with my views on human care as a moral ideal of nursing.

REFERENCES

1. Watson, J. Nursing: The Philosophy and Science of Caring. Boston: Little, Brown, 1979, 9–10.
2. Gadow, S. Existential advocacy as a form of caring: Technology, truth and touch. Paper presented to Research Seminar Series: The Development of Nursing as a Human Science. School of Nursing, University of Colorado Health Sciences Center. Denver, March 31, 1984.
3. American Nurses' Association. Social policy statement. Kansas City, Mo.: American Nurses' Association, 1980.

4. Watson, J. Nursing: The Philosophy and Science of Caring. Boston: Little, Brown, 1979, 9–21.

BIBLIOGRAPHY

American Nurses' Association. Social policy statement. Kansas City, Mo.: American Nurses' Association, 1980.

Gadow, S. Existential advocacy as a form of caring: Technology, truth, and touch. Paper presented to Research Seminar Series: The Development of Nursing as a Human Science. School of Nursing, University of Colorado Health Sciences Center. Denver, March, 1984.

Watson, J. Nursing: The Philosophy and Science of Caring. Boston: Little, Brown, 1979.

10
METHODOLOGY

The methodologies for studying transpersonal caring and developing nursing as a human science and art require a non-traditional view of science and reside in methods that are based upon different assumptions about the:

1. Nature of reality
2. Nature of inquirer–object–subject relationship
3. Nature of truth statements.

Other components of methods that need to be considered for relevance to nursing as a human science have to do with aspects of trustworthiness of the data (for example, truth value or confidence in findings), applicability, consistency and neutrality. Moreover, one has to consider various other dimensions related to method such as quality criterion, theory source, knowledge types, instruments, design, and setting (see Chapter 2).

The methodologies that are relevant for studying my theory can be classified generally as qualitative–naturalistic–phenomenological field methods of inquiry or a combined qualitative–quantitative inquiry versus a quantitative rationalistic method of inquiry as the exclusive method.

The nursing theory of transpersonal nursing (within the framework of qualitative–phenomenological–naturalistic approach) can employ a variety of qualitative, creative methods for exploring meanings of human existence, illness, human caring, and human capacities for healing. These include earlier descriptive approaches such as existential case studies[1] and content analysis; other methods for consideration include ethnomethodology (a phenomenological approach used at the social–cultural level).

A prominent method that is increasingly being acknowledged as an appropriate method for nursing by a number of nursing theorists and researchers is the phenomenological method. Since most of the other qualitative approaches in some way or other are phenomenological it is worthy of further development. This chapter will concentrate on explaining phenomenological research (both descriptive phenomenology and transcendental phenomenology) and applying it to the concept of caring and the human phenomenon of loss–grief. However, it is important to point out that these and other methods are in need of further development and practice, while new creative methods are in need of development as well. It is conceivable to me that literary, poetic, artistic works, for example, may be employed as methods for theorizing about and researching certain human phenomena. Nurses are encouraged to create new approaches that are appropriate for the phenomena under study. For now I confine my work to the specifics of the phenomenological method.

DESCRIPTIVE PHENOMENOLOGICAL METHODOLOGY

This method "consists of describing or explicating experience in the language of experience."[2] The method attempts to describe and understand human experiences as they appear in awareness. These experiences can include such a phenomenon as human care, but also experiences related to human health and illness conditions, such as loss–grieving, anxiety, hope, despair, love, loneliness, spiritual self, higher sense of consciousness, and related human experiences and concepts of existence.

In brief, the subject matter of phenomenological research is human experiences—their types and their structures, along with their subjective meaning, essence, and relationships.

Husserl was concerned that a phenomenological analysis of experiences should not be confused with a psychological analysis of experiences.[3] Psychology is viewed in this sense as an empirical science that studies experiences as empirical events in an empirical world, with all its descriptions and generalizations referring to experiences in this empirical context.

Husserl's[4] idea of phenomenology involves a different attitude: it involves "placing within brackets" the existential, "historical" aspect of experiences and concentrating on the "essence" or the "ideal types" exemplified by the experiences that we either have or are able to conceive of ourselves as having. Phenomenology studies such essences and clarifies the various relationships between them.

Heidegger's[5] point of view was that experiences that merit the greatest philosophical attention are those that find expression in poetry. He believed that only through a searching phenomenological analysis of experiences could we hope to achieve a clarification of the meaning of *being*.

Human phenomena (such as caring and events of being, that is, illness, health) are not objectlike; they cannot be inspected or studied in the manner of objects. They have to do with the "how" rather than the "what." They are not neutral items that call for a neutral and detached independent description. They have to do with modes of existing and the meaning of being. The human phenomena of nursing make themselves known through moods, feelings, and emotions of experiences.

Although there are different views regarding the course that phenomenological research should pursue and the principal objectives of such research, what unites the different views is an acceptance of the general principle that priority should be given to an analysis of experiences from the point of view of those who have the experiences or are able to have them.

Even though there are gradually more and more objections about the traditional, rationalistic, quantitative methodologies for human science, and specifically nursing science, it is still difficult to transform those objections into a concrete research program.

Some of the work of Professor Amedeo Giorgi, of Duquesne University, Pittsburgh,[6] and the research group at the Department of Education at the University of Goteborg, Sweden, have recently elaborated an empirical approach that is phenomenological. From their point of view the contribution of phenomenology as a method is that it offers an alternative way of looking at the investigation of human phenomena.

Merleau-Ponty's ideas also have important considerations for nursing.[7]

According to Merleau-Ponty, the key to understanding phenomenology is to take, as a given, that humans behave but to give human behavior its "proper ontological status," that is, to take it out of the purely physical domain. Merleau-Ponty then conceptualized behavior as a relation between the subject and the world, where the relationship is a dialectic one. Consequently any human phenomena are in a subject–world relation, because embodied in the concept is the idea that that person is necessarily conscious of "something" and thereby directed toward the world. The acknowledgment of the unity of subject–world relationship allows the possibility of our being able to be faithful to the experience of the world as subjectively viewed. We can describe neither the objective nor the subjective world, but only the world as we experience it.

In order to allow for the inseparability of subject–world, it is important to make a distinction between different levels of attending (being in) to the world.[8] At one level we are handling the world in a *prereflective* manner, which means we are close to it, directly involved in it, living it. In Merleau-Ponty's words, "I am outside myself in the world of my project."[9]

However, an experience embodies a dialogue between subject and world and suggests that there is something more than just the world we direct our attention to at the prereflective level. If we are to be able to attend to something else, for example, the *way* of experiencing, then we need a second level of attending called the *reflective level.* Thus, at the reflective level the object of an investigation is the relation between what is experienced (for example, human care) and how it is experienced.[10]

According to Giorgi,[11] Alexandersson,[12] Marton,[13] and others[14] conducting phenomenological research, the key to phenomenology as a method is that we take into account the way the world is experienced; this requires *phenomenological reduction.* It consists of considering not only what is experienced but the mode or manner in which it is experienced.

As Alexandersson stated, "in this reduction we take a step back from the lived experience and reflect upon it—on what is the necessary structure of that experience, on what it must be like in order for those perceptions to occur."[15] According to Richard Zaner,[16] phenomenology's chief method is that of reduction. Phenomenological reduction is explained by him in terms of "shifts of focal attention" and consequent "reflective orientations."

Phenomenological reduction is not at all akin to what has come to be known as reductivism (for example, in philosophy of science); it has nothing whatever to do with any attempt to simplify or economize, much less to try to explain one region by showing it to be reducible to another. Rather, "the basic thrust is found in the literal meaning of the term 'reduction': a leading back to origins, beginnings, which have become obscure, hidden, or covered over by other things."[17]

The phenomenological reduction in no way denies what is naturally believed in or posited by our natural consciousness, but rather is a deliberate effort to "suspend" or "put in abeyance" that attitude to examine it in depth. The reduction, then, is the systematic effort to bring the natural attitude into focus by considering not only what is experienced but how it is experienced.

The phenomenological reduction method is a two-step procedure:

1. The experience is bracketed/suspended/regarded as an appearance.
2. The experience phenomenon is imaginatively varied to obtain the invariant feature of the phenomenon—to discuss the necessary structure of the experience or the *essence* of phenomenon.

The concept of essence here does not mean absolute, definite, and beyond cultural dependencies, but more pragmatically represents the deepest understanding available, established on the basis of intersubjective agreement of a given context. Intersubjectivity consists of having others independently describe a phenomenon as experienced and reported and then comparing the results.

Descriptive-Empirical Phenomenological Research Protocol

This section is based on the recent advances and uses of this method in Sweden. I have freely drawn upon the work of Alexandersson,[18] to which I was exposed in 1981–82 during a postdoctoral Kellogg Fellowship in Australia.

Some approaches for carrying out the descriptive research protocol are:

1. The researcher obtains naive descriptons from other people about their experience of a given phenomenon (puts the experience in brackets and varies imaginately in order to reach the invariant structure of the phenomenon); this also places description in context of subject not investigator.
2. The researcher derives essences using those descriptive and reduction methods, wherein the researcher attempts to understand the various ways an experience has presented itself.
3. The researcher aims to find an objective description of the subjective variations obtained.

An example of a research project that attempts to give concrete form to the theoretical and methological considerations here is a project on loss and caring carried out in Western Australia with tribal Aborigines.

The subjects were asked to describe, in as much detail as possible, a situation in which loss and caring occurred for them. The descriptions thus obtained were analyzed in order to get at the structure of loss and caring in each case. These structures vary between individuals and the aim of every analysis was to reach the meaning of loss and caring in each case.

It was assumed that each meaning would contribute, in its own way, to an even fuller understanding of the phenomena of loss and caring by giving credibility to the phenomena as it is lived. This was accomplished by analzying each description according to the following points:[19]

1. Reading through each protocol to get a sense of the whole.
2. Dividing the protocol into "meaning units" or constituents as expressed by the subject, marking the protocol each time a transition in meaning is perceived. Ask the question "does this say something different from the previous one with respect to the theme?"
3. Interrogating each meaning unit for its psychological–nursing–human care relevance.
4. Reducing meaning units to those that characterize the experience. (Carrying out a "free-imaginative variation" on each one to check that these statements capture the essence of that situation.)
5. Integrating the statements into a structure, synthesizing the statements into an integrated whole that embodies the structure of that

learning experience in such a way that it could be compared with structures derived from other situations.

Although these five steps represent the order in which an analysis is undertaken, they provide a fairly meager impression of the work. In the words of Kris Swanson-Kauffman, a recent doctoral graduate, "It is the rigor and careful analysis of an actual protocol that gives an idea of the exactitude of this work."

In summary, the phenomenological method is a kind of research that is viewed as an alternative method to quantitative rationalistic method but can be complementary to other kinds of research. Its aims are generally description, analysis, and understanding of experiences.

In the search for meaning with this method, "the meaning *is* the measurement."

Since the method of phenomenology aims at description, analysis, and understanding of experiences, it is directed toward experiential description. It is further concerned with the notion of essence, used here to refer to the common intersubjective meaning of human experience of a certain aspect of reality. Phenomenology is also basically methodological in that the findings are substance oriented and would refer to anything that can be said about how people perceive, experience, and conceptualize a given human phenomenon. Lastly, the method is directed toward both the conceptual and the experiential as well as what is thought of about the lived world of the experience. (Phenomenology then is prereflective—living it; reflective—thoughts about how it is lived; and experiential—the way it is experienced.)

(To illustrate the application of this method, two examples of loss-caring phenomena are analyzed using Alexandersson's protocol.)

RAW DATA: PROTOCOL ON CARING AND LOSS

The following is an analysis of a description of loss and caring by a group of tribal Aboriginal men in Cundeelee, Western Australia, an outback Aboriginal settlement, in May, 1982.

Laws have been handed down and we abide by law—going back to Moses; they are handed down from generation to generation.

When I see places my forefathers have been, I imagine myself being there. It tells a story about my ancestors. I feel I am a part of the land itself—the very soil. A sorrowful feeling is someone trying to take the land—destroy our very being inside. We were told not to move one rock or remove anything. We knew uranium was there. We were told (by our Dreamtime) not to come near the place until told to go to the place down there with 1000 other minerals. It's spellbound, harmful, still alive—takes thousands of other minerals to destroy. The Dreamtime (tells us) minerals are there with purpose to keep uranium nonradioactive while in the ground.

Dreamtime told us the Dreamtime man is so powerful that he destroyed everything before him. Birds, trees, animals, water, everything before him.

Man in Dreamtime was put down there by gods. He was destroyed, burnt, and left in their buried. But he didn't actually die. He was in a Big Sleep while in there and he couldn't destroy or hurt anyone and we knew that. We knew that once (land, rocks) uncovered it would unlease the man and it would go on destroying. Dreamtime is true today and it was true then. We abide and we know what causes sickness. It's all been predicted. It has handed down to us to abide by law of land. Very sacred—our thoughts and dreams.

When someone dies in tribe, the closest family relative is to be told first, before the family. It is important that Sister tell the relatives because they won't believe unless it comes from Sister. But important that Sister check with community elders to make sure they inform the correct family.

After someone dies, only inlaws and (aboriginal) community have burial and go to grave. Close family stays together and mourns at home. Wails, groans, moans, fall prostrate on ground, no clothing; bash self in head and body because they feel to blame.

Have first burial (only tribal community) and no close family attends. We (those family closest to person) *do not* go to grave site. We don't want body to be placed there, don't want to see. We share the grief back home. The community goes to grave site and meets, and others from different areas come (during bereavement). People (police) should not intrude because of gathering. Don't interfere. This is time you and I allowed to be there. Should have full time to be in the community. Then second burial happens only when husband/wife/parents (closest family) say they are "ready". It has to be at least a year. It can be up to two years or longer. A person ready (for second burial) when free of hurt, pain, grief, memories; mind will be clear, headaches gone; then when ready to be happy again, he'll be free. That's when second burial occurs. This time called "time out" is necessary to get ready for second burial, to be free to be happy again.

Synthesized Statement

The community needs time off, rest day, time out. The community needs time to be together and go to graveside during bereavement.

This caring continues as nurse shares the pain and shows sympathy. The caring includes not interfering with person/family, not intruding (or not allowing others to intrude) into the community's time. The community needs time off, a rest day, time out to meet others from different areas, time to go to gravesites, and time out during bereavement.

Question: What caring do your people want from nurses during loss (through death)?

First comfort by holding, embracing; that
    Shows she cares
    Share the sorrow
    Letting us mourn, leave us alone
    Let us have time out—time off

We need time out; a rest day—that extends to the community. If Sisters (nurses) go (next) day after and embrace–comfort then people know that Sisters care. When they have time to put aside, then know they care. Share the pain, show sympathy.

APPLICATION OF DESCRIPTIVE-EMPIRICAL METHOD

A Caring Meaning Units Constituents as expressed by S.	B Caring Capturing the Meaning Units that characterize the Experience
First *comfort* by holding, embracing— *Show* care *Share* sorrow, pain *Let* us mourn Leave alone *Let us have time* Nurse have *time* (shows care)— then know Nurse cares Let *community* have time; let community show grief Don't intrude with community Don't interfere Put aside time next day Next day—also comfort— embrace When Nurse has time—shows care—then know.	*Need time*—time off, time out time for community time for family time for Nurse time to put aside to show time (rest day) time alone time allowed to be there (together) full time to be in community Caring Behaviors Caring do's: physical presence, physical acts—holding, embracing, showing, visiting ("go next day") after; sharing (sorrow, pain, grief) exchanges—sharing—physical acts, emotions, griefs, sorrow, time, letting be ("let us mourn"), leaving alone; allowing being; *give* time. Caring don'ts: intrude, don't interfere with community.

A Loss–Grief Meaning units Constituents as expressed by S.	B Loss–Grief Capturing the Meaning units that characterize the experience of grieving
Loss—after someone dies. First burial—only inlaws and community go to grave; close family stays together, mourn at home, wails, groans; fall	1. Close family stays together. 2. Mourn at home. 3. Mourning—wails, groans. 4. Falling prostate on ground. 5. Without clothing.

prostrate on ground; no
clothes; bash self—head, body;
feel to blame.
Second burial—happens only
when close family ready; has
to be at least 1 year; can be 2
years or longer.
Ready (for second burial)—when
person *free* of hurt, pain, grief,
memories; *mind* will be clear,
headaches gone; when person
ready to be happy, he'll be
free; this time called *time-out;*
necessary to get ready for second burial and be "free" to be
happy again.
Grieving person—needs comfort,
needs time, visitation, needs
to have time holding, em-
bracing; needs to mourn with
close family and not be in-
truded; needs to share pain,
sorrow with close friend at
home.

6. Bashing head and body.
7. Mourner feels blame.
8. Only inlaws and community
   go to grave first burial.
9. Second burial—happens only when close family ready.
10. Has to be at least one year.
11. Can be two years or longer.
12. Ready when person free of
    hurt, pain, grief, memories;
    mind will be clear; head-
    aches gone.
13. When ready to be happy,
    person will be free.
14. This time called *time-out*
    and is necessary to be free
    and be happy again.

**Author's synthesis of meaning units and raw data: A specific descrip-
tion of caring wanted by Aboriginal from nurses during loss**

People want comfort by nurses during immediate loss experience. This
comfort is expressed initially by holding and embracing that shows the
nurse cares.

This is followed by nurse sharing sorrow and letting person mourn. Caring then desired includes a request to be left alone and let the person have
time; time-off, a rest day, that time-off extends to include the community (whole tribe in settlement).

In order for the nurse's caring to be known (by the bereaved person),
the nurse must put aside time and give time. This putting aside and giving of time by the nurse is shown by the nurse. The next day (visiting)
and repeating the comfort (giving time, embracing). Then they know the
nurse cares. The nurse needs time for caring to be known. The person
needs time to be alone, to mourn, rest, have time out, time off (to show
grief at home).

Application of Findings to Theory of Transpersonal Caring

The human experience of a significant loss affects a person's being in the
world; loss creates an incongruence between person's self as perceived and
self as experienced. There is disharmony among the three major spheres
of the person's being in the world—mind/emotions–body–soul. All three

spheres—the total person—is affected as the person experiences the profound pain and wound that cuts through to the very soul and alters the sense of self and the phenomenal field. The "I"is not equal to the "me." A major part of a person's being is changed; the sense of self is totally changed and in a condition of disarray. The mind/emotions are affected by the sense of blame, the sorrow, the pain.

The nurse, through transpersonal caring and caring transactions, can enter into the intensely personal-subjective relationship with the grieving person—person to person—by embracing, holding, sharing the sorrow. There is a need for time to be with a person and time to be alone for some inner self work. The nurse through transpersonal caring allows the person to explore and experience their inner fear, blame, sorrow, grief, by allowing privacy, allowing time, allowing community to express its public grief while family has private grief at home.

The caring nurse does not allow herself or others (for example, police) to intrude or interfere in the grieving time-out process. He or she allows the person to live his or her feelings, perceptions, and experiences—as he or she reflects them back to the person and allows them to be released in their own way consonant with phenomenal field and past life history. The nurse also appropriately gains knowledge about the subjective personal meanings of the experience so he or she can better allow them to be expressed in ways that are meaningful and right for the individual—not the nurse.

The nurse also experiences his or her own feelings and has his or her own personal world of meaning to explore and allow his or her self "to live" the experiences, the feelings, and the perceptions and accept these as part of the changing, developing, self, and as way to be more fully human.

The nurse allows for caring transactions to arise from an actual caring occasion of loss, but in turn that process becomes part of the life history of each person and part of some deeper, larger, complex pattern of life. As such, the nurse can reach out and touch the spiritual senses or the very hurt soul of a grieving person.

When a person is experiencing profound loss, their very being-essence-soul is troubled and needs to be tended and nourished.

SUMMARY

An empirical-descriptive phenomenological analysis of human experiences in health and illness can provide a rich description of human meanings of experiences as lived by a person. As such the experiences can lead to increased understanding of human behavior in health and illness and to means of exploring the human process of caring in nursing. Human experiences "cannot be measured or experimented with—they are simply there and can only be explicated in their givenness."[20]

In order to explore and study the concepts inherent within the deeply human process I have called transpersonal caring, one has to acknowledge that the ideas cannot be studied by a simplistic reduction associated with traditional quantitative research methods. Complex theories associated with humans, life, phenomenal fields, and other such entities require that researchers grow beyond the traditional methods and explore differing approaches more compatible with the human art and science of nursing.

An empirical–descriptive phenomenological orientation in theory and practice and phenomenological methodology for nursing research is one approach that requires additional development, further use and refinement. As an orientation and method, phenomenology holds promise for nursing science in that it can complement traditional approaches and accomplish its own contributions toward development of human care knowledge and advancement of the human art and science of nursing.

TRANSCENDENTAL OR DEPTH PHENOMENOLOGY AND POETIC RESULTS

One further development of empirical phenomenology is the method referred to as transcendental or depth phenomenology.* The next section elaborates upon the extension of empirical phenomenology to the notion of transcendental–poetic expression of phenomenology. The section concludes with excerpts of transcendental poetry that capture the same experience as the empirical approach but present it in very different language.

Some recent thinking in phenomenology takes the position that phenomenology is not just a descriptive methodology but a poetic formulation of language that attempts to interpret experiential evidence and dynamics.[21] The notion of transcendental phenomenology has been described as "a powerful guardian of the dream we name, echoing an earlier renaissance, the dream of humanism."[23]

The humanism in phenomenology consists in the capacity of this experiential methodology to be a method of self-awareness and self-understanding that contributes "not only to our satisfaction but can guide us toward a well-being that really fulfills our human nature."[24] Transcendental phenomenology is concerned with the very *depth* of experience and an *openness* to our nature, our potential for being. Thus, in being true to *depth*, *openness*, and *humanism*, transcendental phenomenology requires an experiential depth and is indeed transcendental (of pure facts and descriptions) insofar as it "cherishes the process of deepening and opening and nurtures with methodological guidance a continuing movement of *self-transcendence.*"[25]

---

* This section is drawn from the work of David Levin (1983).[22]

Husserl realized that the transcendental method gives access to a hidden or deep realm of experience that functions according to inwrought principles of its own order, without being obliged to rectify it as an objective-factual thought, but rather to consider the deeply rich experiential process as a form of transcendence from the experience itself.[26]

The transcendental notion in this sense acknowledges "the inexhaustible depth and openness of the implicit treasury of human experiences."[27] According to Merleau-Ponty, the transcendental method is not a means of establishing an autonomous transcendental subjectivity, but rather a gesture that initiates "the perpetual beginning of reflection, at the point where the individual life begins to reflect on itself."[28]

In this sense then, transcendental or depth phenomenology is very closely aligned with art and science in that it is the act of bringing truth of an experience into being. As such, transcendental phenomenology is an almost perfect methodological match for studying and developing nursing as a human science and art, and researching the human care process described as transpersonal caring.

In numerous instances in this work, I have attempted to make a case for why and how the traditional quantitative methods were inconsistent with nursing subject matter and premises related to nursing being a human science and art. In this last section, I am taking that position a step further. Based upon my philosophy, and my view of science, humans, and nursing, it is also necessary to abandon the original Husserlian vision of phenomenology being a "rigorous phenomenological science" and the paradigm that the method is "pure description." According to Merleau-Ponty, Levin, and others, to envision phenomenology as pure description places it within the outdated paradigm of rationalism and logical positivism rather than within an approach that is true to the vital and creative dimension of depth experience,[29] indeed, an approach that achieves a "poetic" effect in that the articulations of the experience, as felt and lived, transcend the facts and pure description of the experiences. As such, transcendental phenomenology allows for some poetic ambiguity, some sensuous resonance, characteristic of an experiential approach to study, but a conception of description that helps us to focus on our experience and bring it into expression.

The methodological analysis and procedures, adopted by Giorgi,[30] Alexandersson,[31] and Marton[32] presented in the previous section, as innovative and nontraditional as they may seem, are an example of the more traditional orthodox phenomenological approach to experience. However, Merleau-Ponty, Levin, and others allow for the phenomenologist researcher to go beyond the surface phenomenology and relate the description to the depth and openness of experience; this method allows for reflection upon the experience, the process of emergent meaning, insight, and the actual expression of the experience takes into account the dynamic movement involved in the process of reflecting and using language. The result is or can be poetic. The result unites science with art.

As the researcher allows for the level of depth or transcendence of the description of the experience, to contact another dimension of one's being, which taps depth and openness, there is potential for being oneself and being true to oneself. As we allow this to occur, however, we find ourselves in touch with much more than the fact, the pure description, and in touch with much more of our being than we can know in a cognitive, intellectual way.

Whenever such transcendence occurs the descriptions of the phenomenological experience never truly fit the experience; there remains an elusiveness and an ambiguity in the meanings. However, the transcendent data as presented have the potential to allow the researcher to be true to his or her self and acknowledge the openness of the process. In order to be true to one's self and the depth humanism in which phenomenology is embedded, one must acknowledge the openness of the process and also stay with it in "truthful harmony, because openness to our being is the center, the heart of our essential nature."[33]

To further quote Levin,[34] "if there be any truth, then, in the trancendental method (of phenomenology), it must be that the transcendental is not just a method for understanding the facticity of experience; but that it is also a way of enjoying, or appreciating the intrinsically creative and open nature of experiences, because appreciation of this nature is a necessary condition for true and authentic existential knowledge."

Therefore, if phenomenology is to be true to the human science and art of nursing and the human care process, it must penetrate beneath the surface of familiar, habitually organized, and standardized experience. What is at issue is presenting the phenomenological experience; in the use of language, as such, the language must have a transcendental and no longer a mundane relationship to our experience. The process of transcendental phenomenology commits us to a language that encourages existential authenticity. Moreover, authentic speech that is true to experience, and also true to our deepest experience of expression is languaging that touches and opens up the transformative process.[35]

According to Levin, "authentic languaging, trascendentally gets involved in the potential for growth implied in our reflection-upon-experience." This difference is not so much a question of their different contexts of meaning as it is a question of their way of relating to the experiential process. The transcendental reduction does not change the meaning of our words; rather, it changes how our words relate to the experience.[36] For example, an Aboriginal does not own land, but the land owns him. It's an entirely different relationship, one to another.

Therefore, in acknowledging the transcendental nature of both the human experience of phenomenology as well as the nature of its expression, it is necessary to acknowledge that the way in which experience is expressed is at least as important as the content, the facts, and the pure description of the experience.

In other words how could cold, unfeeling, totally detached dogmatic words and tone possibly teach the truth or deep meaning of a human phenomenon associated with human care, transpersonal caring and grief, and convey experiences of great sorrow, great beauty, passion, and joy.[37] We cannot convey the need for compassion, complexity, or for cultivating feeling and sensibility in words that are bereft of warmth, kindness, and good feeling.[38]

In order for phenomenology to be true to humanism of experience, the language used to describe the experience must penetrate beneath the factual surface of everyday experience and allow us to see anew, to address a powerful truth that moves us to realize our deepest tendencies and existential meaningfulness.[39] (. . . to allow for release of feelings one has been longing and wishing to express.)

In the ideas of Heidegger,[40] the result is poetizing (Dichtung), which is the true vocation of the experiential phenomenologist. According to Levin again, poetizing, in this instance, is necessary in that transcendental depth phenomenology if focused and reflective of depth human experiences, cannot be other than poetic.[41] This occurs first, by expressing and conveying—embodying—the beautifully good feeling that spontaneously arises with the saying of that which is true to experience; and second, by virtue of this truth (being-in-truth), disclosing a more open space. Poetizing addresses and lays claim to our potential for being, and, like a metaphor, it carries us forwards.

To continue in Levin's words[42]:

> Any experiential articulation that (1) is *rooted* in a truly felt experience, (2) *emerges* from that experience in a felt movement of self-expression, and (3) *maintains its contact* with the original, spontaneous thrust of experience, even in the phase of completed expression, will at least *tend* to be (tend to *sound*) poetic.

He goes on to conclude that, "in phenomenological discourse, therefore, the deepest transcendental truth of an existentially authentic languaging of experience will be articulated with the sensuous resonance, the emotional spaciousness, and the elemental openness of the poetic word."

Heidegger[43] referred to the importance of actually undergoing an experience with language, and letting our experiences speak for itself. He indicated we should let ourselves be transformed by our participation in the process.[44] Moreover, poetic expression has the power to touch and move us, to open and transport us. Thus, the poetic quality is related to the experiential meaning and, indeed, deepens the meaning, the felt senses, so that there is increased openness to describe and preserve the truth and depth of the experience.

Indeed, when a phenomenologist is true to the depths of the moving human experience, he or she almost naturally poetizes. Levin suggests that poetizing descriptions of transcendental phenomenology serves as visualizations, imaginative projections. As such, the power of imagination through poetizing actually brings us nearer to a way of being (we are already living) and moves us deeper into that open transcendental realm of experience.

If we are to consider this deep level of phenomenology, that is beyond pure descriptions, and allow for the transcendental experience as both felt and expressed through poetic language, then we have to give up the correspondence theory of truth and adapt the *aletheia* theory of truth which is associated with discovery of the unknown, or unconcealment.[45] We also have to abandon the notion of a "descriptive" as well as "factual," "quantitative" truth that simply corresponds to facts and figures. The human science and art of nursing and human care, which is indeed transpersonal, must incorporate feeling, depth of experience, and transcendental processes that result in poetic expression; that moves us toward authentic experiential expression and helps us maintain openness with our humanism and our potential for growth. Finally, transcendental phenomenology also reminds us that as nurses, in either practice or research efforts, we are first of all human beings, capable of transcending the moment, capable of engaging in a truly felt experience, emerging from that experience with a desire for self-expression while still being capable of maintaining contact with the original, spontaneous thrust of the experience.

Example of Transcendental Phenomenology and Poetic Expression

As a way of contrasting the phenomenological analysis of pure description with transcendental phenomenology, I am concluding with excerpts of poetry written immediately following my phenomenological research experience with an Aboriginal tribe in Western Australia. One can go back and read the "description" of the loss and caring experiences reported earlier (pages 87–88) and now realize the limited expression of the described empirical experience, compared with the transcendental poetic expression of the experience.

Incidentally, the writing of the poetry was in itself a transcendent experience for me. As I left the Aboriginal mission and the other friends in Kalgoorlie, Western Australia I was faced with an overnight train ride back to civilization in Perth, the coastal city. During the train ride I had an overwhelming desire to express my feelings and capture the gestalt of my experience, and also capture the gestalt of the Aborigines' loss-caring experiences. In reflecting back on my field notes, the data, and the entire experience, the poetic expression formulated on the overnight train ride captures the truth of the experience and the meaning of the human phenomena better than any of the factual data that are described without any feeling of personal involvement. The transcendental description was rooted in a truly felt experience that indeed transcended the here and now and resulted in a much more comprehensive expression of the deeply human experiential process.

DREAMTIME AND SHARING THE TEARS
WITH WONGI TRIBE OF CUNDEELEE

Jean Watson
Western Australia, May 1982

An arch of eyebrows
that represents the bush
of the Bush Country he comes from,
has left and is now longing to return to.
With his sugar bag he is preparing
to return home
    to his people
before the end and after
    the reburial.
The "time out" of grief is
approaching two years or longer.
When he's ready he'll let
    his people know.
His dark skin, so dark
around his eyes
I have to look two or three times
to catch the shine of the brown eyes
    that know all,
        see all, yet are.
He speaks about visions, the Milky Way,
the black hole,
they told him in dreamtime.

His people have known
    for thousands of years.
The end will come when
the black hole is in the
Milky Way and the Emu
    in the stars
makes a drumming noise.
The heavens will open at
    the black hole.
You can see the religious awakening
happening all over the world.

There's that side of Bill
and then there's the power of the
    Spirit
I felt in the dark
of Cundeelee Camp.
We floundered and wondered
how it would be to talk—
    whether he would "see" and
I could be real with him.

And we spoke
   and left it—
Spoke and
   left it.
I wandered to be alone in the
   dark of Cundeelee.
The steps outside
the sister's compound.
The ashes of the coals
The fire of Mangrove Root
Chanting children
Singing of God
"If you're happy and you know it
clap your hands."
He stepped up to me
I felt it.
   The warm flow of
his acceptance
his approval
his readiness
   to be with me.
It came from my throat
   and chest
and moved to him
   with warmth
and enclosed me.
The words didn't matter.
   I couldn't hear.
But we both knew it was
   okay.
But others kept coming, talking
   noises and words.
Time, Patience
The group gathering—
   the Elders
   red head bands
The eldest with cowboy hat.

The women and children
   naked and dressed.
The dingoes all gathered
   to fill the soul
with singing, and clapping
and telling their woes.
The sins, the drinking, the
   tearing their souls.
The visions, the hurts,
   the finding the Lord.
The arrows that pointed
   from the clouds above

and led them to Christ,
    the Son, the Lord,
and now they had come
    to spread the Word.

All questioning and listening
and singing their songs.
Bill came up and told me
so specially so "I want you
    to have some of my kangaroo tail."
I floundered, but nodded
I was pleased to accept, not
knowing what ritual presented
    itself.

After watching Bill peel it
    of skin and fur, I reckoned
    I ate it, instead of observe.
So, little by little with proper
    bites,
I daintily abided his
    appetite.
Gourmet?, not really—but
    surely not bad.
To taste a little oily
    kangaroo tail.
But then came the photos
to catch it all live—cause
no one would believe me
back in American eyes.

Struggling and trying to
    find my way.
Do I look out to the
    people
    or hide in their ways?
Bill taught me that hiding the eyes
may be better for them
    to capture the wholeness
    without rude chagrin.
And then came the Elders.
The men of the tribe who
    asked for Bill's counsel
to help them decide what
someone from Boulder could
    possibly do with people as
    remote as
Aborigines from Cundeelee.

He tried to express
    with his best Wongi tongue

"That lady's a Sister,
  a nurse;
  a bloke if you will
who seeketh the wisdom
  of you and me.
  The feelings of people from Cundeelee."

The caring and loss concepts
  of Aboriginal disgrace
so beautifully spelt, but so
  ineptly expressed.
Bill had the words, the signs and
the grace
to explain it all kindly
  to the full men without haste.
He used words of the
  Wongi
  that left me behind
but carried the men to
  Dreamtime Beyond.
His white head of wisdom,
  he knew beyond all—
  the right way to cross over
  the lands that bind—
  the hearts that twine
  when the worlds unfold.
  I found myself stumbling
  when they knew it all.

They said it like this.
It's a world behind in Dreamtime.
it's finished
  we're free
after proper "time out" to
  clear the head—
of headaches, the memories,
the sorrow and grief.
The Wongi for Caring
  means "sharing the tears"
But after the comfort,
which includes an embrace,
that only then tells us
  You care for our race.

We also want help
  to get to our nearest place
to find our relatives
to carry on the debate
And how to release
  the Community
  from the State

The State that denies them
   the "time out" they need
   to fill up their sorrow
   pour out the grief

The greed of the Country,
   the business, the mines
That won't let the people
   wail as they need
When all they ask for
   is time out to grieve.

It's so deep, so painful
They say it all hurts
They must be alone to
   properly mourn
Even if it includes
   some self abuse
That white man can't conger
   so obtuse
Cause our fields are
   from Dreamtime
   that penetrates years
and guides all our people
   to bury their fears
   to trust one another
   and learn to obey
   to believe in our brothers
   in spite of their ways.
To wish for them goodness
   even when they pray
   regardless of memories
and haunting decay.

They sit and they hope
they sleep in the night
beyond all comprehension
   of white man's likes.

They watch in the heavens
for the Milky Way paths
   Emu noises and black holes
   That quake.
The earth, The water, The birds
   and The fowl,
The animals and trees
and certainly the tail of
   The lowliest Kangaroo
   that shivers and quails.
When all men are gone
   according to scale.

We knew it all through hundreds
   of stories. Dreamtime continues to wail.
The grief you say is only one tale

The forefather's Dreamtimes
   are mighty and powerful
   and all we entail.
The earth is our partner
our part of this life
It's sacred to feel
   and sinful to fail.

The Dreamtime shows us
   how never to err
When it comes to uncovering
   The rocks and the soil,
Because of the death and
   destruction that's left
   in the trail.
Instead of objects and gemstones
   and people with dreams
But only death and destruction
   Youla—I mean
Even the name change
   to *Yolaria*
   Can't change the time
   That's played for Mankind
   once the notes are produced.

So leave it alone, don't tamper
   with fate
The Uranium's a hate of the
   whole human race
Except the miners and sharers
   of great
The Dreamers of Visions
for money and fame;
   for rights and lights and
   claims are at stake.

But listen to wisdom to time
   and my stories of late
That have been told by the
   Dreamtime and hold us
   awake
If only we hear The children
   and chants in the night
The stars in the heavens
   That show us what's right
The soul's reawakening will
   come in the night

If you listen to Dreamtime
before all goes quiet.

Wongi reminders will haunt
after all
When Cundeelee mission
has nowhere to fall.

REFERENCES

1. Binswanger, L. Insanity as life-historical phenomenon and as mental disease: The case of Ilse. In R. May, E. Angel, & H.F. Ellenberger, (Eds.), Existence. New York: Basic Books, 1958, 214–236.
2. Hall, C.S. & Lindzey, F.: Theories of Personality, (3rd ed.). New York: Wiley, 1978, 332.
3. Pivcevic, E. (Ed.). Phenomenology and Philosophical Understanding. London: Cambridge University Press, 1975, 271–286.
4. Husserl, E. Phenomenological Psychology. The Hague: Martinus Nijhoff, 1977, 20–45.
5. Heidegger, M. Being and Time. New York: Harper & Row, Pub., 1962, 205.
6. Giorgi, A. An application of phenomenological method in psychology. In A. Giorgi, C. Fisher, & E. Murray, (Eds.). Duquesne Studies in Phenomenological Psychology, (Vol. 2). Pittsburgh, Pa.: Duquesne University Press, 1975, 82–104.
7. Merleau-Ponty, M. The Primacy of Perception. Evanston, Ill.: Northwestern University Press, 1964, 186.
8. Alexandersson, C. Amedeo Giorgi's Empirical Phenomenology. (Publication No. 3). Swedish Council for Research in Humanities and Social Sciences, Department of Education, University of Goteborg, Sweden, 1981, **3**, 1–35.
9. Merleau-Ponty, M.: The Primacy of Perception. 186, 187.
10. Ibid.
11. Giorgi, A. Psychology as a Human Science: A Phenomenologically Based Approach. New York: Harper & Row, Pub., 1970.
12. Alexandersson, C. Amedeo Giorgi's Empirical Phenomenology, 1–35.
13. Marton, F. Phenomenography—Describing conceptions of the world around us. Instructional Science, 1981, **10**, 177–200.
14. Zaner, R.: On the sense of method of phenomenology. In E. Pivcevic, (Ed.), Phenomenology and Philosophical Understanding. London: Cambridge University Press, 1975, 125–140.
15. Alexandersson, C. Amedeo Giorgi's Empirical Phenomenology, 1–35.
16. Zaner, R. On the sense of method in phenomenology. In E. Pivcevic, (Ed.), Phenomenology and Philosophical Understanding, 125.
17. Ibid., 126.
18. Alexandersson, C. Amedeo Giorgi's Empirical Phenomenology, 1–35.
19. Ibid., 18.
20. Van Kaam, A. Existential Foundations of Psychology, (Vol. 3), Pittsburgh, Pa.: Duquesne University Press, 1966, 187.

21. Levin, D. The poetic function in phenomenological discourse. In W. McBride & C. Schrag, (Eds.), Phenomenology in a Pluralistic Context. Albany, N.Y.: State University of New York Press, 1983, 216-234.
22. Ibid.
23. Ibid., 217.
24. Ibid.
25. Ibid., 218.
26. Ibid., 218, 219.
27. Ibid., 219.
28. Merleau-Ponty, M. Phenomenology of Perception. London: Routledge & Kegan, 1962.
29. Levin, D. The poetic function in phenomenological discourse. In W. McBride & C. Schrag, (Eds.), Phenomenology in a Pluralistic Context, 216-234.
30. Giorgi, A. An application of phenomenological method in psychology. In A. Giorgi, C. Fisher & E. Murray, (Eds.), Duquesne Studies in Phenomenological Psychology. 82-104.
31. Alexandersson, C., Amedeo Giorgi's Empirical Phenomenology, 1-35.
32. Marton, E. Phenomenography—Describing conceptions of the world around us, 177-200.
33. Levin, D. The poetic function in phenomenological discourse, 221.
34. Ibid.
35. Ibid.
36. Ibid.
37. Ibid.
38. Ibid., 228.
39. Ibid., 221.
40. Heidegger, M. Poetry, Language and Thought. New York: Harper & Row, Pub., 1975, 155.
41. Levin, D., 228.
42. Ibid., 228, 229.
43. Heidegger, M. The nature of language. In M. Heidegger (Ed.), On The Way to Language. New York: Harper & Row, Pub. 1971, 98.
44. Heidegger, M. The anaximander fragment. In M. Heidegger (Ed.), Early Greek Thinking. New York: Harper & Row, Pub., 1975, 55.
45. Marton, F., Phenomenography—Describing conceptions of the world around us, 177-200.

BIBLIOGRAPHY

Alexandersson, C. Amedeo Giorgi's Empirical Phenomenology. (Publication No. 3). Swedish Council for Research in Humanities and Social Sciences, Department of Education, University of Goteborg, Sweden, 1981.

Barret, W. Irrational Man. New York: Doubleday, 1962.

Betteridge, H.T. (Ed.). The New Cassell's German Dictionary. New York: Funk and Wagnalls, 1958.

Binswanger, L. Insanity as life-historical phenomenon and as mental disease: The case of Ilse. In R. May, E. Angel & H.I. Ellenberger, (Eds.), Existence. New York: Basic Books, 1958.

Boss, M. Psychoanalysis and Daseinsanalysis. New York: Basic Books, 1963.

Buber, M. I and Thou, (2nd ed.). New York: Scribners, 1958.

Capra, F. The Turning Point. New York: Simon & Schuster, 1982.

Davis, A.J. The phenomenological approach in nursing research. In N. Chaska, (Ed.), The Nursing Profession: Views Through the Mist. New York: McGraw-Hill, 1978.

Dennis, N. Personal communication and health seminar, Western Australian Institute of Technology, Australia, 1982.

Frankl, V.E. Man's Search for Meaning. New York: Washington Square Press, 1963.

Frye, N. The Educated Imagination. Bloomington, Ind.: Indiana University Press, 1964.

Giorgi, A. An application of phenomenological method in psychology. In A. Giorgi, C. Fisher & E. Murray, (Eds.), Duquesne Studies in Phenomenological Psychology, (Vol. 2). Pittsburgh, Pa.: Duquesne University Press, 1975.

Giorgi, A. Psychology as a Human Science: A Phenomenologically Based Approach. New York: Harper & Row, Pub. 1970.

Hall, C.S. & Lindzey, F. Theories of Personality, (3rd ed.). New York: Wiley, 1978.

Heelan, P. Hermeneutics of experimental science in the context of life-world. In D. Ihde & R. Zaner, (Eds.), Interdisciplinary Phenomenology. The Hague, Netherlands: Martinus Nijhoff, 1977.

Heidegger, M. The anaximander fragment. In M. Heidegger (Ed.), Early Greek Thinking. New York: Harper & Row, Pub., 1975, 55.

Heidegger, M. Being and Time. New York: Harper & Row, Pub., 1962.

Heidegger, M. The nature of language. In M. Heidegger (Ed.), On The Way to Language. New York: Harper & Row, Pub., 1975.

Heidegger, M. Poetry, Language and Thought. New York: Harper & Row, Pub., 1975.

Hora, T. Transcendence and healing. Journal of Existential Psychiatry, 1961, 1, 501.

Husserl, E. The Crisis of European Sciences and Transcendental Phenomenology. Evanston Ill.: Northwestern University Press, 1970.

Husserl, E. Phenomenological Psychology. The Hague: Martinus Nijhoff, 1977.

Ihde, D. Existential Technics. Albany, N.Y.: State University of New York Press, 1983.

Ihde, D. & Zaney, R. (Ed.). Interdisciplinary Phenomenology. The Hague: Martinus Nijhoff, 1977.

Johnson, R.E. In Quest of a New Psychology. New York, Human Sciences Press, 1975.

Koch, S. Psychology and emerging concepts of science as unitary. In T. Wann, (Ed.), Behaviorism and Phenomenology: Contrasting Basis for Modern Psychology. Chicago, Ill.: University of Chicago Press, 1964.

Kohler, W. Gestalt Psychology: An Introduction to New Concepts in Psychology. New York: Liveright, 1947.

Levin, D. The poetic function in phenomenological discourse. In W. McBride & C. Schrag, (Eds.), Phenomenology in a Pluralistic Context. Albany, N.Y.: State University of New York Press, 1983.

Lewin, K. A Dynamic Theory of Personality. New York: McGraw-Hill, 1935.

Marton, F. Phenomenography—Describing conceptions of the world around us. Instructional Science, 1981, **10**, 177-200.

Maslow, A.H. Toward a Psychology of Being, (2nd ed.). Princeton, N.J.: Van Nostrand, 1968.

McBride, W.L. & Schrag, C.O. (Eds.). Phenomenology in a Pluralistic Context. Albany N.Y.: State University of New York Press, 1983.

Merleau-Ponty, M. Phenomenology of Perception. London: Routledge & Kegan, 1962.

Merleau-Ponty, M. The Primacy of Perception. Evanston, Ill.: Northwestern University Press, 1964.

Mohanty, J.N. The destiny of transcendental philosophy. In W.L. McBride & C.O. Schrag (Eds.). Phenomenology in a Pluralistic Context. Albany, N.Y.: State University of New York Press, 1983.

Munhall, P.L. Nursing philosophy and nursing research: In apposition or opposition? Nursing Research, 1982, **31**. 176, 177, 181.

Oiler, C. The phenomenological approach in nursing research. Nursing Research, 1982, **31**, 178-181.

Omery, A. Phenomenology: A method for nursing research. Advances in Nursing Science, 1982 **5**(2), 49-63.

Pivcevic, E. (Ed.). Phenomenology and Philosophical Understanding. London: Cambridge University Press, 1975.

Psathas, G. Ethnomethodology as a phenomenological approach in the social sciences. In D. Ihde & R. Zaner (Eds.), Interdisciplinary Phenomenology. The Hague: Martinus-Nijhoff, 1977.

Psathas, G. Phenomenological Sociology: Issues and Applications. New York: Wiley, 1973.

Sartre, J. Being and Nothingness. New York: Philosophical Library, 1956.

Spiegelberg, H. On some human uses of phenomenology. In F. J. Smith, (Ed.) Phenomenology in Perspective. The Hague: Martinus Nijhoff, 1970.

Spiegelberg, H. The Phenomenological Movement, (Vol. 2). The Hague: Martinus-Nijhoff, 1965.

Straus, E. Phenomenological Psychology. New York: Basic Books, 1966.

Tillich, P. The Courage To Be. New Haven, Conn.: Yale University Press, 1952.

Tolstoy, L. The Wisdom of Tolstoy. New York: Philosophical Library, 1968. (Translated by Huntington Smith as an abridgement of L. Tolstoy, My Religion. London: Walter Scott Publ., 1889.)

Valle, R.S. & King, M. (Eds.). Existential Phenomenological Alternatives for Psychology. New York: Oxford University Press, 1978.

Van Kaam, A. Existential Foundations of Psychology, (Vol. 3). Pittsburgh, Pa.: Duquesne University Press, 1966.

Van Kaam, A. Phenomenological analysis: Exemplified by a study of the experience of being really understood. Individual Psychology, 1959, **15**, 66-72.

Vaught, C.G. The Quest for Wholeness. Albany, N.Y.: State University of New York Press, 1983.

Watson, J. Nursing: The Philosophy and Science of Caring. Boston: Little, Brown, 1979.

Watson, J. Professional identity crisis—Is nursing finally growing up? American Journal of Nursing, 1981, 2, 1488–1490.

Watson, J. Supporting materials for and introduction to the new undergraduate curriculum. Unpublished. University of Colorado School of Nursing, 1976.

Yalom, I.D. The Theory and Practice of Group Psychotherapy, (2nd ed.). New York: Basic Books, 1975.

Zaner, R. On the sense of method in phenomenology. In E. Pivcevic, (Ed.), Phenomenology and Philosophical Understanding. London: Cambridge University Press, 1975.

Zubek, J.P. (Ed.). Sensory Deprivation. New York: Appleton-Century-Crofts, 1969.

INDEX

Tables are indicated by *t*; notes are indicated by *n*.

109190

RT 84.5 .W367 1988
Watson, Jean, 1940—
Nursing

DATE DUE